B

Progress in Mathematics
Volume 78

Series Editors
J. Oesterlé
A. Weinstein

W. Borho J.-L. Brylinski R. MacPherson

Nilpotent Orbits, Primitive Ideals, and Characteristic Classes

A Geometric Perspective in Ring Theory

1989

Birkhäuser
Boston · Basel · Berlin

W. Borho
BUGH - FB 7
Gaußstraße 20
5600 Wuppertal 1
Federal Republic of Germany

J -L. Brylinski
Department of Mathematics
McAllister 305
Pennsylvania State University
University Park, Pennsylvania
U.S.A.

R. MacPherson
Department of Mathematics
Massachusetts Institute of Technology
Cambridge, Massachusetts
U.S.A.

Library of Congress Cataloguing-in-Publication Data
Borho, W. (Walter), 1945-
Nilpotent orbits, primitive ideals, and characteristics classes : a
geometric perspective in ring theory / W. Borho, J.-L. Brylinski, R.
MacPherson.
 p. cm. — (Progress in mathematics ; v. 78)
 Includes bibliographical references.

 1. Rings (Algebra) I. Brylinski, J.L. (Jean-Luc)
II.MacPherson, Robert, 1944- . III. Title. IV. Series: Progress
in mathematics (Boston, Mass.) ; vol. 78.
QA247.B65 1989
512'.4—dc20

Dr. Brylinski's work is partially supported by NSF Grant DMS-8815774.
Dr. MacPherson's work is partially supported by NSF Grant DMS-8803083.

Printed on acid-free paper.

ISBN-13: 978-1-4612-8910-4 e-ISBN-13: 978-1-4612-4558-2
DOI: 10.1007/978-1-4612-4558-2

Text prepared by the authors in camera-ready form.

9 8 7 6 5 4 3 2 1

CONTENTS

INTRODUCTION

1. The Subject Matter.

Consider a complex semisimple Lie group G with Lie algebra **g** and Weyl group W. In this book, we present a geometric perspective on the following circle of ideas:

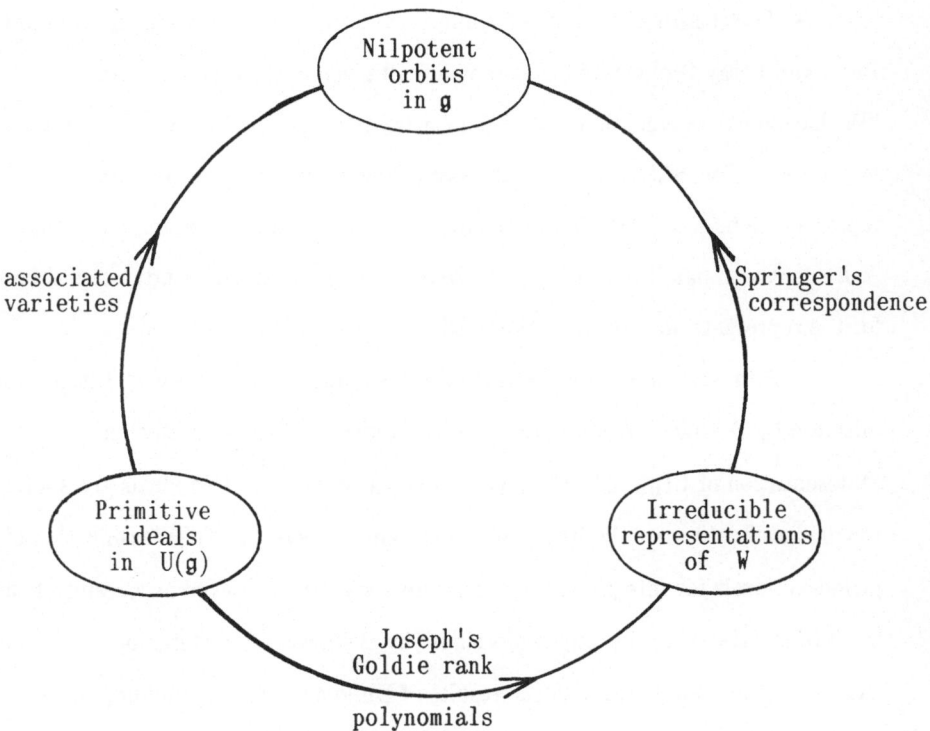

The "vertices" of this graph are some of the most important objects in representation theory. Each has a theory in its own right, and each has had its own independent historical development.

 — A nilpotent orbit is an orbit of the adjoint action of G on **g** which contains the zero element of **g** in its closure. (For the special linear group

$G = SL(n,C)$, whose Lie algebra \mathfrak{g} is all n × n matrices with trace zero, an adjoint orbit consists of all matrices with a given Jordan canonical form; such an orbit is nilpotent if the Jordan form has only zeros on the diagonal. In this case, the nilpotent orbits are classified by partitions of n, given by the sizes of the Jordan blocks.) The closures of the nilpotent orbits are singular in general, and understanding their singularities is an important problem.

— The classification of irreducible Weyl group representations is quite old. Borel and Leray [Bo] showed that they are all given by certain spaces of "W–harmonic" polynomials on \mathfrak{t}, the Cartan subalgebra of \mathfrak{g}, with the natural W action. (For $SL(n,C)$, W is the symmetric group on n letters; its representations were classified by Young — they correspond to Young tableaux, or equivalently to partitions of n.) A more modern question is to study natural bases for the representation spaces of the irreducible Weyl group representations.

— Primitive ideals are the newest of the three. Their systematic study was initiated by Dixmier. A primitive ideal is the kernel of some irreducible representation of $U(\mathfrak{g})$. Therefore the classification of primitive ideals gives a first rough classification for the irreducible representations of \mathfrak{g}. The classification of primitive ideals for integral central characters was first achieved as a result of the proof of the Kazhdan–Lusztig conjectures (by Beilinson–Bernstein and Brylinski–Kashiwara) through the results of Duflo and Joseph (building on work of Jantzen, Barbash, Vogan, and others).

The "edges" of the graph are connections which have been the subject of recent research.

— The Springer correspondence associates to an irreducible W representation a nilpotent orbit together with a local system over it.

— Any ideal in $U(\mathfrak{g})$ has an associated variety which is a subvariety of \mathfrak{g}. The associated variety of a primitive ideal is the closure of a single nilpotent orbit,

as was first conjectured by Borho and Jantzen [Bo].

— Joseph associated to a primitive ideal a W—harmonic polynomial on t,
which is a basis element of one of the irreducible Weyl group representations.

2. Our Approach

The original approaches to the connections represented by the edges of the
graph above were somewhat algebraic in nature. Our approach here will be to use
geometry wherever possible. In particular, we make use of the geometric methods of
equivariant K—theory, \mathcal{D}—modules, and intersection homology.

For both the connections between nilpotent orbits and representation theory
(the Springer correspondence and the associated varieties), the fact has clearly
emerged, through the work of many authors, that nilpotent orbits do not carry the
full picture. One has to consider the geometry of the moment map: $\pi : T^*X \to \mathfrak{g}$,
where X is the variety of complete flags for G (i.e. $X = G/B$, for B a Borel
subgroup of G), and T^*X is its cotangent bundle. It is easy to describe π if one
views $T^*(X)$ as the variety of pairs (b',ξ), where b′ is a Borel subalgebra of \mathfrak{g},
and $\xi \in [b',b']$. Then $\pi(b',\xi) = \xi$.

For \mathcal{O} a nilpotent orbit, its inverse image $\pi^{-1}(\mathcal{O})$ under π has several
irreducible components; the closure of each component \mathcal{K} is a cone bundle over X
included in T^*X. We call these cone bundles orbital cone bundles. They will be
the main geometric object that we study. An orbital cone bundle is a homogeneous
bundle $\mathcal{K} = G \times^B C$, for C a cone in the fiber u of T^*X at the base point (which
is equal to the nilpotent radical of b). Such cone bundles are prominent in
Springer's theory of Weyl group representations where they correspond to natural
basis elements. Also, the characteristic variety of the \mathcal{D}—module corresponding to a
primitive ideal is a union of orbital cone bundles, and its projection in \mathfrak{g} is the
associated variety.

We study here <u>characteristic classes</u> of orbital cone bundles K over X, both in ordinary cohomology and in equivariant cohomology. The characteristic class of K is the element of the cohomology of X defined by the cohomological restriction to the zero section of T^*X of the fundamental class $[K]$ of K. The characteristic classes in ordinary cohomology may be computed from the Segre classes of K, introduced by Fulton and MacPherson [FM]. By identifying the cohomology of X with a space of W–harmonic polynomials on t, the characteristic classes of orbital cone bundles allow a direct construction of the Springer representation associated to a nilpotent orbit O in terms of W–harmonic polynomials on t. One may also similarly define a characteristic class in the Grothendieck group K(X). The relation between the Segre classes and the class in K–theory is described in Chapter 1 of this book.

In the case of a homogeneous cone bundle K over X, there is also a class in equivariant cohomology (or rather, a class in equivariant K–theory $K_G(X)$); this class is obtained again by restricting to X, the zero–section of T^*X, the class $[O_K]$ in $K_G(T^*X)$, for $K \subset T^*X$ a cone bundle and O_K its structural sheaf. One also obtains such a class in equivariant K–theory, associated to any equivariant sheaf on T^*X, and therefore to any equivariant D–module on X. All these classes become concrete polynomials on the Cartan subalgebra t or the Cartan subgroup T, and $K_G(X)$ is the representation ring R(T) of T.

The element of R(T) thus associated with a D–module M on X appears to be a sort of "formal character"; in the case where M is the localization of $U(\mathfrak{g})/I$, for I a primitive ideal, one finds essentially the formal character of a corresponding highest weight module. This gives a geometric interpretation of Joseph's theory relating characters of highest weight modules to the Goldie rank polynomials of their annihilator ideals. Together with previous work of the authors [BB3], it gives

a geometric formulation of the classification of primitive ideals, with integral central characters.

To a large extent, this work builds on the previous work of the authors, more precisely the work of the first and third author on nilpotent orbits and Weyl group representations (using methods from intersection homology) **[BM]**, and the work of the first and second author on primitive ideals and their characteristic varieties (using \mathcal{D}–modules) **[BB3]**. We also assume some background in algebraic groups (in Chapters 1 and 4), in topology (in Chapter 1), in representation theory (Chapter 5). On the other hand, we give a self–contained introduction to equivariant K–theory in the algebraic case. Also, we use very little non–commutative ring theory. We give full statements or at least detailed references for information we borrow from other sources.

We hope that the book will be readable to readers with varied interests and backgrounds; the main purpose of the book is to present a general geometric picture, which intends to unify various aspects of representation theory and the theory of algebraic groups. If certain proofs appear too technical, it is possible to omit them without losing sight of the general picture.

3. Contents of the Book.

We now give a description of the contents of the five chapters. In Chapter 1, we first review Segre classes of cone bundles (in the sense of **[FM]**), and we compare them with a characteristic class in K–theory. We then recall (1.5 and 1.6) Borel's description of the cohomology of the flag variety X, and in particular the realization of $H^{2d}(X)$ as the space of W–harmonic polynomials on t of degree d. We compute the Segre class of some homogeneous vector bundles. In 1.7, we introduce the homogeneous cone bundles $\mathcal{K} = G \times^B C$ over X, where C is a

component of a fiber of the moment map. This is the central geometric object considered in the book.

In 1.8, we state the main result of Chapter 1, which says that for any nilpotent orbit \mathcal{O}, if we denote by \mathcal{K}_i the irreducible components of $\overline{\pi^{-1}(\mathcal{O})}$, the characteristic classes of the \mathcal{K}_i form a basis of the Springer representation of W associated with \mathcal{O}. This theorem is reformulated in 1.9, as the commutativity of a certain diagram, involving intersection homology. The proof is given in 1–10 to 1–12. It makes heavy use of the results of [BM]. In 1–14, we compare this to Springer's original construction of Weyl group representations. In 1–15, we deduce Hotta's transformation formulae, as a corollary of Theorem 1.8.

Chapter 2 is an introduction to equivariant K–theory in the context of algebraic geometry. In 2.9, we discuss the restriction map to the zero section $i^*:$ $K_G(E) \rightarrow K_G(X)$, for E a G–bundle over X. According to Thomason [T1], this restriction map is always an isomorphism. In 2.10, we recall Grothendieck's γ–filtration, which is important in later chapters. In 2.12, we explain how a (weakly) G–equivariant \mathcal{D}–module over Y gives rise to a class in $K_G(T^*Y)$.

Chapter 3 studies the equivariant K–theory $K_T(E)$, where T is an algebraic torus which operates linearly on the vector space E, in such a way that all weights of T in E are positive. For any finitely generated T–equivariant $S(E^*)$–module F, one defines its formal character $ch(F)$, which is of the form $\Delta(E^*)^{-1} \sum_{\lambda \in \Lambda} a_\lambda e^\lambda$, where Λ is the group of characters of T, and $\Delta(E^*) = \prod_{\mu \in S} (1 - e^\mu)$ (if S is the set– with multiplicities– of weights of T in E).

In 3.7, we show that $ch(F)\Delta(E^*)$, as an element of $R(T)$, is simply $i^*[F]$, where $[F] \in K_T(X)$ and $i : \{0\} \rightarrow E$. In 3.8, we conclude that the γ–degree of $[F]$ is equal to the degree of the lowest order term in the power series $ch(F)\Delta(E^*)$.

This lowest order term q_F is a homogeneous polynomial on t of degree d (3.9). Theorem 3.10 states that d is also equal to the codimension in E of the support of F. The proof is given in 3.11–3.13. In 3.15, we show how Hilbert–Samuel polynomials are obtained, for the special case $T = G_m$. In 3.16, we prove that q_F is characterized by its restriction to one–parameter subgroups of T. In 3.17 and 3.18, we use this fact to give another proof of Theorem 3.10.

In Chapter 4, we describe explicitly $K_G(T^*X)$ for $X = G/B$ the flag variety. It is isomorphic to $K_G(X)$, hence to $R(T)$ (4.3). In 4.4, we deduce from Chapter 3 that for \mathcal{F} a G–equivariant coherent sheaf on T^*X, the γ–degree of $[\mathcal{F}] \in K_G(T^*X)$ is the codimension of the support of \mathcal{F}. In 4.5, we define the equivariant characteristic class $Q_G(\mathcal{F})$, and we compute it from a formal character. In 4.6, we explain the relation of $Q_G(\mathcal{F})$ with the characteristic class $Q(\mathcal{F})$ of Chapter 1. Theorem 1 in 4.7 states that for $\mathcal{K} = G \times^B C$ an orbital cone bundle in T^*X, $Q_G(\mathcal{K})$ is a harmonic polynomial. We draw some consequences in 4.7. In 4.8, we show that this class $Q_G(\mathcal{K})$ coincides with Joseph's characteristic polynomial p_C.

We reprove the transformation formula of Joseph for his polynomials, under some simple symmetry in W, by an adaptation of his methods to our geometric context. This is done in 4.10–4.12. From this the theorem follows (4.13). In 4.14, we deduce Hotta's transformation formula. In 4.15, we give the explicit computation of $Q_G(\mathcal{K})$, when C is a complete intersection in \mathbf{u}. We discuss a possible generalization of this method in 4.17.

In Chapter 5, we apply the constructions and results of Chapter 4 to Harish–Chandra bimodules and primitive ideals. In 5.1, we define the characteristic class $P(M)$ of a $U(\mathbf{g})$–module M as $P(M) := Q(\text{gr } \mathcal{M})$, where \mathcal{M} is the D–module obtained by localizing X (in the sense of [BeBe] and [BKa]). In 5.3, we recall results of [BB3], mostly the fact that the characteristic variety $\text{Ch}(H^\ell)$ of a

Harish–Chandra bimodule H, viewed as a left U(\mathfrak{g})–module, is a union of orbital cone bundles. In 5.4, we show more precisely that if H is <u>homogeneous</u>, Ch(H$^\ell$) is a union of orbital cone bundles, all of the same dimension. We recall that the correspondence of [BeGe] associates to H a highest weight module L. We show (Theorem 5.5) that P(H$^\ell$) is equal to p(L), where p(L) is Joseph's character polynomial.

In 5.7, we show that p(L) is a harmonic polynomial. In 5.8, we introduce the equivariant class P$_G$(H), and show that P(H$^\ell$) is the harmonic part of P$_G$(H). In 5.10, we recall the notion of <u>Goldie rank</u> of a module over a ring, and of uniform module over a ring. We show that for H a simple Harish–Chandra bimodule, U \subset H$^\ell$ a uniform submodule, <u>Ch</u>(H$^\ell$) is equal to rk(H$^\ell$) <u>Ch</u>(U) (<u>Ch</u> denotes a characteristic cycle, which is an algebraic cycle in T*X). This means that the Goldie rank is measured by the characteristic cycle. In 5.12, we define, for any w \in W, a polynomial P$_w$ as P(U), for U a uniform submodule of a Harish–Chandra bimodule associated with w. In 5.13, we show that for any w, P$_w$ generates an irreducible W–module.

In 5.14, we use this to give a new proof of the irreducibility of associated varieties of primitive ideals. In 5.15, we compute the characteristic class of a Harish–Chandra bimodule in terms of Kazhdan–Lusztig polynomials. In 5.16, we obtain Joseph's formula for the Goldie rank [J3]. In 5.17, we derive the factorization formula of King [Ki].

The results obtained in Chapter 5 are essentially due to Joseph [J3],[Ja]. The originality of our approach is that we substitute \mathcal{D}–modules and equivariant K–theory for non–commutative algebra. The relevance of equivariant K–theory in representation theory has been independently demonstrated by the work of Kazhdan and Lusztig [KL2], [KL3] and of Ginsburg [Gi2], [Gi3] on representations of p–adic groups and Hecke algebras. Let us mention also that Vogan proposed in [V2] a

restoration of the orbit method of Dixmier and Kirillov. This was further studied in [MG]. Further deep relations between primitive ideals and nilpotent orbits appear in recent work of Joseph [J5], Hotta–Kashiwara [H3], and Barbasch–Vogan [BV3]. Clearly, this whole area is going to be very active for several years to come.

4. Acknowledgements

We are grateful to several institutions for their hospitality and support, without which the gatherings of the authors necessary to produce this work would have been impossible. These include the MPI für Mathematik in Bonn, the RIMS in Kyoto, the UBC in Vancouver, the Università di Roma I in Rome, the IHES in Bûres–sur–Yvette, the CIRM in Marseilles, Brown University in Providence, and UBGH in Wuppertal.

In addition, we are grateful to the NSF in Washington, the CNRS in Gif–sur–Yvette, and the MPG in München for financial support. The first author benefitted from visits to the Weizman Institute in Rehovot and the Universities of Warwick and Leeds.

We had very useful conversations with Bill Fulton about the material in Chapter 1 and with Klaus Bongartz about the material in Chapter 3. We owe a tremendous intellectual debt to the work of Tony Joseph, whose beautiful ideas provided the main stimulus for our work.

Finally, we would like to thank Anne Richard, whose final typescript you are now reading.

The research reported on here was announced in [BBM2]. The material of Chapter 1 appeared in [BBM3]. Earlier versions of Chapters 2–4 were preprinted at the IHES (March, 1986) and at the BUGH Wuppertal (September 1986).

§1. A DESCRIPTION OF SPRINGER'S WEYL GROUP REPRESENTATIONS IN TERMS OF CHARACTERISTIC CLASSES OF CONE BUNDLES.

Our "algebraic varieties" are reduced, and defined over an algebraically closed base field k of characteristic 0. We restrict attention to the case $k = \mathbb{C}$ whenever we find it convenient for topological interpretations. If not otherwise stated, we consider de Rham (co)–homology etc. with coefficients in k.

1.1 Segre classes of cone bundles [FM], [Fu].

A *cone* in a vector space is a union of lines through the origin. Let Y be a non–singular algebraic variety. A *cone bundle* over Y is an algebraic variety K equipped with 1) an action of the multiplicative group k^* (or G_m), and 2) a morphism $K \to Y$ making K a fibre bundle over Y, which admits a closed embedding into a vector bundle E over Y, respecting both data 1) and 2). (In particular, each fibre of K is embedded into the corresponding fibre of E as a closed cone.)

Each cone bundle K over Y determines a certain "*characteristic class*" s(K) in the cohomology ring $H^*(Y)$ ($= H^*(Y,k)$), called its *Segre class*. It can be axiomatically defined by the following two properties:

(1) The Segre class of a vector bundle E over Y is the inverse of its total Chern class c(E), i.e.

$$s(E) = c(E)^{-1} .$$

(2) The Segre class is compatible with proper pushforwards in the following sense: Given a commutative square of two cone bundles

with f proper and \tilde{f} proper birational and G_m-equivariant, the functorial ring homomorphism $f_*{:}H^*(Y') \to H^*(Y)$ maps $s(K')$ to $s(K)$.

We refer to Fulton–MacPherson [FM], or Fulton [Fu] chapter 4, for the proof of existence and uniqueness of such classes $s(K)$ (even on singular base spaces Y), as well as for more properties and historical background.

1.2 Characteristic class of a subvariety of a vector bundle.

Let E be a vector bundle over the non–singular equidimensional variety Y. Then each subvariety \mathcal{V} of codimension d in E determines a cohomology class of degree 2d on the base space Y, denoted $Q(\mathcal{V}) \in H^{2d}(Y)$, and defined as follows. We start here with a geometric definition for the case $k = \mathbb{C}$, and postpone the statement of a more general, formal algebraic definition to the next section (1.3). If m resp. n are the dimension of the fibre resp. base of E, the real dimension of \mathcal{V} is 2m+2n–2d. It therefore defines a canonical homology class $[\mathcal{V}]$ in the closed support homology group $H^{cl}_{2m+2n-2d}(E)$. Homology with closed supports is sometimes called Borel–Moore homology. It is the homology of the complex of locally finite singular chains. A representative for $[\mathcal{V}]$ can be obtained by triangulating \mathcal{V} (with infinitely many simplices). Since E is non–singular, Poincaré duality gives a canonical identification of $H^{cl}_{2m+2n-2d}(E)$ with the cohomology group $H^{2d}(E)$. Since the zero section $\sigma : Y \to E$ is a homotopy equivalence, we have an induced graded ring isomorphism $\sigma^* : H^*(E) \xrightarrow{\sim} H^*(Y)$, which we call "homological intersection with the zero section." Then, by definition, the characteristic class $Q(\mathcal{V})$ is obtained from the canonical class $[\mathcal{V}]$ by Poincaré duality followed by homological intersection with the zero section.

For the case of a cone bundle $\mathcal{V} = K$, there is the following nice formula for this characteristic class $Q(K)$ in terms of Chern and Segre classes, due to Fulton-MacPherson [FM], see also [Fu], p. 73, 4.1.8.

Proposition: *If the subvariety* $V \subset E$ *is a cone bundle* $K = V$ *over* Y *of codimension* d *in* E, *then*

$$Q(K) = [c(E) \cdot s(K)]^{2d} .$$

Here $[...]^i$ means homogeneous part in degree i.

1.3 Characteristic class determined by a sheaf on a bundle.

The following definition, generalizing 1.2, will not be used until chapter 3. It may serve as an algebraic alternative to 1.2, but its main purpose is to link up the results of the present chapter with our work in later chapters. For general background, we refer to [SGA6], [Ma], [Fu], [FL], [BFM].

With notations as in 1.2, we consider now an arbitrary coherent sheaf **F** of \mathcal{O}_E–modules whose support supp(**F**) has codimension d in E, and we define a characteristic class Q(**F**) determined by **F** in $H^{2d}(Y)$ as follows. We consider the Grothendieck ring K(E) of (the category of) all coherent \mathcal{O}_E–modules, filtered by the subgroups $K^j(E)$ (j $\in \mathbb{Z}$) generated by coherent \mathcal{O}_E–modules whose support has codimension \geq j. The ring structure on K(E) comes from its identification with the Grothendieck ring of locally free \mathcal{O}_E–modules (cf. 1.4). We note that this filtration coincides with Grothendieck's γ–filtration at least after tensoring with \mathbb{Q}, see [FL], p. 182, Proposition 5.5; but note here and in the sequel our convention that our coefficients are tacitly extended to k, for convenience. We next apply the Chern character ch:K(E) → $H^*(E)$, which is a functorial homomorphism of the Grothendieck ring into the cohomology ring [Fu], [FL], [Hi]. Finally, we intersect homologically with the zero–section σ: Y → E (as in 1.2), and take the degree 2d homogeneous part. In summary, we define

(2) $$Q(\mathbf{F}) := [\sigma^* \mathrm{ch}[\mathbf{F}]]^{2d} .$$

We may sometimes also consider the "total" class attached to **F** by

(3) $$Q_{\mathrm{total}}(\mathbf{F}) := \sigma^* \mathrm{ch}[\mathbf{F}] .$$

Remarks. a) Note that we could reverse the order of σ^* and ch in this definition, since functoriality of the Chern character provides a commutative diagram

$$
\begin{array}{ccc}
K(E) & \xrightarrow{\ \sigma^*\ } & K(Y) \\
\text{ch} \downarrow & & \text{ch} \downarrow \\
H^*(E) & \xrightarrow{\ \sigma^*\ } & H^*(Y) \ .
\end{array}
$$

b) Let us point out that $Q_{total}(F)$ has degree $\geq 2d$, or in other words: $Q(F)$ is either the lowest degree term of this class, or else zero. To see this, recall that the ring $K(E)$ is filtered by the Grothendieck subgroups $K^j(E)$ of coherent \mathcal{O}_E–modules with support of codimension $\geq j$. Then the Chern character respects this filtration up to a doubling of degrees, that is

$$
\text{ch } K^d(E) \subset \bigoplus_{j \geq 2d} H^j(E) \ ,
$$

which implies our claim.

The Chern character thus induces a homomorphism

$$
\text{gr ch} : \text{gr } K(E) \to H^*(E)
$$

of the *associated graded ring* gr $K(E)$ into cohomology, given by

(4) $\qquad \text{gr}_j \text{ch} : \text{gr}_j K(E) := K^j(E)/K^{j+1}(E) \to H^{2j}(E)$, for all degrees j.

1.4 Comparison of the two definitions for Q.

In order to establish compatibility of the topological definition for $Q(\mathcal{V})$ in 1.2 with the algebraic definition in 1.3, we have to check that in case $k = \mathbb{C}$, the class $Q(\mathcal{O}_{\mathcal{V}})$ of 1.3 coincides with $Q(\mathcal{V})$ of 1.2. For the comparison, the algebraic analogue of Borel–Moore homology of E is the Grothendieck group $K_0(E)$ of coherent \mathcal{O}_E–modules, while the cohomology of E corresponds to the Grothendieck ring of algebraic vector–bundles (or locally–free sheaves), denoted $K^0(E)$. Since we are assuming Y (hence E) non–singular, the canonical ("Poincaré duality") map

$K^0(E) \to K_0(E)$ is an isomorphism, which justified the notation $K(E)$ for both groups; but for the present specific purpose, the notational distinction will contribute to clarity. We look at the following diagram

$$
\begin{array}{ccccc}
K_0(E) & \overset{\sim}{\underset{\text{can.}}{\longleftarrow}} & K^0(E) & \overset{\overset{*}{\sigma}}{\underset{\sim}{\longrightarrow}} & K^0(Y) \\
\downarrow{\tau} & \text{can.} & \downarrow{\text{ch}} & & \downarrow{\text{ch}} \\
H_*^{c\ell}(E) & \overset{\sim}{\longleftarrow} & H^*(E) & \overset{\overset{*}{\sigma}}{\underset{\sim}{\longrightarrow}} & H^*(Y)
\end{array}
$$

where τ is the "Todd character" defined in [BFM], which makes the left hand square commutative only up to multiplication by an element in $H_*^{c\ell}(E)$ which is the image under "can." of the Todd class $\mathscr{T}\!d(T_E)$ of the tangent bundle of the smooth variety E. However, this multiplication does never affect the top degree term, i.e., in other words, the above diagram induces a *commutative* one on the associated graded level:

$$
\begin{array}{ccccc}
\mathrm{gr}_d K_0(E) & \overset{\sim}{\longleftarrow} & \mathrm{gr}_d K^0(E) & \overset{\sim}{\longrightarrow} & \mathrm{gr}_d K^0(Y) \\
\downarrow & & \downarrow & & \downarrow \\
H_{2m+2n-2d}^{c\ell} & \overset{\sim}{\longleftarrow} & H^{2d}(E) & \overset{\sim}{\longrightarrow} & H^{2d}(Y)
\end{array}
$$

Now the desired equality follows from the fact that the canonical class $[\mathcal{O}_\mathcal{V}]$ mod $K_0^d(E)$ determined by \mathcal{V} in $\mathrm{gr}_d K_0(E) = K_0^d(E)/K_0^{d+1}(E)$ maps onto the canonical class $[\mathcal{V}]$ determined by \mathcal{V} in $H_{2m+2n-2d}^{c\ell}(E)$.

From the considerations above, we may draw a more general conclusion, which will become significant in chapter 3. Recall that one may attach to each irreducible component \mathcal{V} of the support $\mathrm{supp}(F)$ of our coherent \mathcal{O}_E–module a well–defined integer multiplicity $m_\mathcal{V}(\underline{F}) \geq 0$. Hence we may define an algebraic cycle (supporting F) as the formal linear combination

$$\mathrm{supp}(F) = \sum_\mathcal{V} m_\mathcal{V}(F)[\mathcal{V}] \quad .$$

Proposition: $Q(F) = \sum_{\mathrm{codim}\,\mathcal{V}=d} m_\mathcal{V}(F)Q(\mathcal{V}) \quad .$

In particular, the characteristic class $Q(F)$ of a coherent \mathcal{O}_E–module F, is

completely determined by its supporting cycle.

Let $F \neq 0$. We claim that

$$[F] - \sum_{\text{codim } \mathcal{V} = d} m_\mathcal{V}(F)[\mathcal{O}_\mathcal{V}] \in K_o^{d+1}(E)$$

has degree strictly larger than $[F]$. Then the proposition will be clear from the preceding discussion. Let S denote the (reduced) support of F, and $\mathcal{V}_1, \ldots, \mathcal{V}_r$ its irreducible components. By an easy dévissage, one may assume that F is an \mathcal{O}_S–module, so we may speak of the class $[F]$ in $K_o(S)$. In each \mathcal{V}_i there exists an open subset $U_i \neq 0$ such that the restriction of F to U_i is a free module of rank $m_{\mathcal{V}_i}(F)$. This means that the element $[F] - \sum_{1 \le i \le r} m_{\mathcal{V}_i}(F)[\mathcal{O}_{\mathcal{V}_i}]$ of $K_o(S)$ restricts to zero in $K_o(U)$, where $U = U_1 \cup \ldots \cup U_r$. By the localization exact sequence for K_o, this element belongs to the image of $K_o(S \backslash U)$ in $K_o(S)$. Since $S \backslash U$ has codimension $\ge d+1$ in E, this implies our claim.

1.5 Homology of the flag variety [Bo], [Hℓ]

In the present paper, we shall apply the previously defined notions in the special case where $Y = X$ is a flag variety and $E = T^*X$ its cotangent bundle. So let us introduce now some of the fundamental notations and facts for this particular situation. We consider a *semisimple*, connected, linear *algebraic group* G defined over k, and fix a *Borel subgroup* B. Then we may define the *flag variety* as $X = G/B$, a complete homogeneous space. Let $n = \dim_k X$. We also fix a maximal torus $T \subset B$ and denote by W the *Weyl group* $W = N_G(T)/T$. For each $w \in W$, the *Schubert cell* X_w is the locally closed subvariety BwB/B, isomorphic to an affine space of dimension $\ell(w)$, the *length* of w with respect to the system of simple reflections determined by B. Since the various Schubert cells X_w ($w \in W$) form a paving of X by affine spaces of even real dimensions $(\dim_R X_w = 2\ell(w))$, they provide a vector space basis for the homology groups:

$$H_{2j}(X) = \bigoplus_{\ell(w)=j} k[X_w] \quad \text{for } j = 0, \ldots, n \ .$$

Let us point out that the existence of such an affine paving causes the Chern character to be an isomorphism $K(X) \stackrel{\sim}{\to} H^*(X)$, so that the two diagrams in 1.4 consist entirely of isomorphisms in the present situation, viz.

$$
\begin{array}{ccccc}
K_0(T^*X) & \stackrel{\sim}{\longleftarrow} & K^0(T^*X) & \stackrel{\sigma^*}{\underset{\sim}{\longrightarrow}} & K^0(X) \\
\tau \downarrow \cong & & \text{ch} \downarrow \cong & & {}_* \text{ch} \downarrow \cong \\
H_*^{c\ell}(T^*X) & \stackrel{\sim}{\longleftarrow} & H^*(T^*X) & \stackrel{\sigma_*}{\underset{\sim}{\longrightarrow}} & H^*(X)
\end{array}
$$

As another special feature of the flag variety case let us mention here as a side remark, that the multiplier $\mathscr{I}d(T_{T^*X}) = 1$ is trivial in this particular case, so that even the left hand square of the above diagram is commutative (cf. the remark below).

Let U be the unipotent radical of B. We denote $\mathfrak{g}, \mathfrak{b}, \mathfrak{t}, \mathfrak{u}$ the Lie algebras of G,B,T,U. For convenience, we sometimes identify \mathfrak{g} with its dual \mathfrak{g}^* via the Killing form. We also identify therefore \mathfrak{u} with the cotangent space of $X = G/B$ at the base point $x = \{B\}$. Then T^*X identifies with the associated fiberbundle $G \times^B \mathfrak{u}$ (as usual, [BM1], [BB] I,III). It is an iterated extension of the line bundles with fibre $\mathfrak{u}^\alpha, \alpha \in \Phi^+$. Here $\Phi^+ \subset \mathfrak{t}^*$ denotes the system of positive roots relative B, and $\mathfrak{u}^\alpha = \mathfrak{g}^\alpha$ is the root space belonging to $\alpha \in \Phi^+$.

Remark: Let us prove that $\mathscr{I}d(T_{T^*X}) = 1$. It suffices to prove that $\mathscr{I}d(T_X)\mathscr{I}d(T_X^*) = 1$. The filtration of \mathfrak{g} by the B–invariant subspaces: $0 \subset \mathfrak{u} \subset \mathfrak{b} \subset \mathfrak{g}$ induces a filtration of $G \times^B \mathfrak{g}$ by homogeneous sub–bundles. Since the whole bundle $G \times^B \mathfrak{g}$ resp. the bundle subquotient $G \times^B (\mathfrak{b}/\mathfrak{u})$ are trivial vector bundles (the latter since B acts trivially on $\mathfrak{b}/\mathfrak{u}$, the former since the B–action extends to a G–action)), we obtain

$$1 = \mathscr{I}d(G \times^B \mathfrak{g}) = \mathscr{I}d(G \times^B \mathfrak{u}) \ \mathscr{I}d(G \times^B (\mathfrak{b}/\mathfrak{u}))\mathscr{I}d(G \times^B (\mathfrak{g}/\mathfrak{b})) = \mathscr{I}d(T_X^*)\mathscr{I}d(T_X),$$

since \mathfrak{u} resp. $\mathfrak{g}/\mathfrak{b} \cong \mathfrak{u}^*$ identify with the cotangent resp. tangent space of X at B. (This sort of phenomenon for the flag variety was first observed in [Mr].)

1.6 Cohomology of the flag variety.

There is a nice explicit description, due to Borel, of the full ring structure of $H^*(X)$, in terms of polynomial functions on \mathfrak{t}, which we shall explain in more detail in the adequate context in chapter 3. Let us state here only basic facts relevant for the present chapter: First, the cohomology of X is trivial in all odd degrees (by the remarks made in 1.5). Second, the cohomology group $H^{2d}(X)$ in some even degree 2d identifies canonically with the vector space $S^d(\mathfrak{t}^*)$ of polynomial functions on \mathfrak{t} which are W–*harmonic*, and homogeneous of degree d, for all $d = 0,1,...,n$. In particular, this provides a linear W action on $H^{2d}(X)$. Third, the cup product of cohomology classes is given by multiplying polynomials modulo the ideal generated by the W–invariants without constant term (cf. 3. or [Bo]).

Convention. From now on, we shall consider the characteristic classes $Q(\mathcal{V})$ resp. $Q(F)$, defined by a subvariety $\mathcal{V} \subset T^*X$ or a coherent \mathcal{O}_{T^*X} –module F as in 1.3 resp. 1.4, as harmonic polynomials on the Cartan subalgebra \mathfrak{t}, if this is convenient; if the difference between a characteristic cohomology class on X and the corresponding *harmonic polynomials* on \mathfrak{t} should matter, then we may refer to the latter as a "characteristic harmonic polynomial."

Example. Let us illustrate the computation of classes $Q(\mathcal{K})$ for cone bundles \mathcal{K} in T^*X, using the conventions above and the formula of 1.2. We consider the special case where the conical fibre $\mathfrak{k} \subset \mathfrak{u}$ is actually a vector space, spanned by root spaces. Let Ψ resp. Ψ' denote the set of positive roots α such that \mathfrak{u}^α *is* resp. *is*

not contained in \mathfrak{k}, so $\mathfrak{k} = \underset{\alpha\in\Psi}{\oplus}\ \mathfrak{g}^\alpha$, and $\mathfrak{k}' = \underset{\alpha\in\Psi}{\oplus}\ , \mathfrak{g}^\alpha$ is a complement of \mathfrak{k} in

u. Then clearly, the codimension of \mathcal{K} in T^*X is given by $d = \dim \mathfrak{k}' = \#\ \Psi'$.

The total Chern classes are given by:

$$c(T^*X) = \underset{\alpha\in\Phi^+}{\Pi}\ (1+\alpha)\ ,\quad c(\mathcal{K}) = \underset{\alpha\in\Psi}{\Pi}\ (1+\alpha) = s(\mathcal{K})^{-1}\ .$$

Hence we may compute the product (cf. 1.2):

$$c(T^*X)s(\mathcal{K}) = c(T^*X)/c(\mathcal{K}) = \underset{\alpha\in\Psi'}{\Pi}\ (1+\alpha)$$

which has highest term $\Pi\alpha$, $\alpha\in\Psi'$, of degree 2d (if each α is given degree 2). Then

the formula in 1.2 says that

$$Q(\mathcal{K}) = \underset{\alpha\in\Psi'}{\Pi}\ \alpha\ .\qquad\qquad(*)$$

Strictly speaking, by our convention above, one has always to take only the

harmonic parts of these products. However, in the case when, e.g. Ψ' is a

parabolic subset, then the product (*) is always harmonic.

1.7 Orbital cone bundles on the flag variety ([BB] III, Appendix B)

We denote by $\pi\colon T^*X \to \mathcal{N}$ the so—called Springer resolution, which is a

G—equivariant proper algebraic map of T^*X onto the cone \mathcal{N} of all nilpotent

elements in \mathfrak{g}. We obtain some particularly nice cone bundles on X by taking the

preimage of a "nilpotent orbit." To be more precise, let $u \in \mathcal{N}$ denote a nilpotent

element, and $\mathcal{O} = \mathcal{O}_u \subset \mathcal{N}$ the G—orbit generated by u under the adjoint action of

G on its Lie algebra. Since a nilpotent orbit is obviously stable under

multiplication by a nonzero scalar, its closure $\overline{\mathcal{O}}$ is clearly a cone, and so is

$\overline{\mathcal{O} \cap u}$, by the same reason. Now consider the preimage $\pi^{-1}\mathcal{O}$ of the orbit in

T^*X. Interpreting T^*X as the associated fibre bundle $G \times^B u$ (notation 1.5), we

have obviously

$$\pi^{-1}\mathcal{O} = G \times^B (\mathcal{O}\cap u)\ ,$$

hence the description of its closure by

$$\overline{\pi^{-1}\mathcal{O}} = G \times^B \overline{\mathcal{O} \cap u}$$

which exhibits its structure as a cone bundle on $G/B = X$ with fibre $\overline{\mathcal{O} \cap u}$.

Moreover, if C_1,\dots,C_r denote the irreducible components of $\overline{\mathcal{O} \cap u}$, then their

associated fibre bundles $\quad \mathcal{K}_i = G \times^B C_i$, for $i = 1,\dots,r$

are the irreducible components of $\overline{\pi^{-1}\mathcal{O}}$. We call these cones C_i resp. cone

bundles \mathcal{K}_i "*orbital* for \mathcal{O}"; any cone in u resp. cone bundle in T^*X is called

"*orbital* ", if it is orbital for some nilpotent orbit (which is then necessarily uniquely

determined). By a result of Spaltenstein [Sp2], all C_i $(1 \le i \le r)$ above have the

same dimension, and hence all \mathcal{K}_i $(1 \le i \le r)$ have the same dimension. From

Steinberg [St], it follows that the common codimension of these cone bundles \mathcal{K}_i in

T^*X is given by

$$d = d_u := \dim \pi^{-1}u = \tfrac{1}{2} \operatorname{codim}_{\mathcal{N}} \mathcal{O} = \operatorname{codim}_{T^*X} \mathcal{K}_i .$$

1.8 Another realization of Springer's Weyl group representation

We are now ready to state the main result of this chapter.

Theorem: *Let $\mathcal{K}_1,\dots,\mathcal{K}_r$ be the orbital cone bundles for some nilpotent orbit \mathcal{O}_u of*
codimension $2d_u$ in \mathcal{N} (as in 1.7). Let $Q(\mathcal{K}_1),\dots,Q(\mathcal{K}_r)$ be their characteristic
classes on X, as defined in 1.2. Then:

a) *These classes are linearly independent.*

b) *They span an irreducible Weyl group submodule of $H^{2d_u}(X)$.*

c) *This irreducible representation is equivalent to Springer's*
representation ρ_u.

For the last statement, we have to recall that Springer [S1] constructs an

irreducible representation of the Weyl group W on $H^{2d_u}(\pi^{-1}u)^{G_u}$, the

G_u–invariants of the top cohomology group of the fibre of his resolution, where G_u

is the isotropy group in G of the nilpotent element u. We denote this representation by $\rho_u = \rho_{(u,1)}$, following our conventions in [BM] 1,2 and [BB] I,III, which differ from Springer's by a sign–character (cf.1.13).

Before going into the proof $(1.9 - 1.12)$, let us point out some immediate useful consequences:

Corollary 1. *The set of classes* $Q(\mathcal{K})$, *where* \mathcal{K} *runs through all orbital cone bundles* (for all nilpotent orbits), *is linearly independent.*

Proof: This follows from a), b), c) of the theorem in combination with the fact that the Springer representations for different nilpotent orbits are pairwise non–equivalent. Q.e.d.

Remark: Note that in a general context, our topological invariant $Q(\mathcal{V})$ defined in 1.2 may often be zero, and hence not of much interest. But for the study of nilpotent orbits, the theorem establishes the usefulness of our concept. Here is another illustration of its use.

Corollary 2: *Let* \mathcal{V} *be the (left) characteristic variety of a primitive ideal in the enveloping algebra* $U(\mathfrak{g})$, *or of a Harish–Chandra bimodule* (cf. [BB] III). *Then* $Q(\mathcal{V}) \neq 0$.

Proof: By loc.cit., \mathcal{V} is a union of orbital cone bundles. Let $\mathcal{K}_1,...,\mathcal{K}_r$ be those of maximal dimension. Then $Q(\mathcal{V}) = Q(\mathcal{K}_1)+...+Q(\mathcal{K}_r) \neq 0$ by corollary 1. Q.e.d.

Let us also mention that the theorem will allow us to understand and reprove Hotta's results [Ho] in a more natural way, see §3.

1.9 Reformulation of the theorem using intersection homology.

To put the theorem into a more formal language, note that the fundamental classes $[\mathcal{K}_1],...,[\mathcal{K}_r]$ of our orbital cone bundles form a basis of the top homology group of $\overline{\pi^{-1}\mathcal{O}_u}$, and that our characteristic class construction (1.2) provides a group homomorphism

$$H^{d}_{4n-2d_u}(\pi^{-1}\mathcal{O}_u) \xrightarrow{\ \delta\ } H^{2d_u}(X) \ ,$$

sending $[\mathcal{K}_i]$ to $Q(\mathcal{K}_i)$. Now part a) of the theorem asserts that δ is *injective*. We shall first reinterpret this map (see the proposition below) in terms of the intersection homology approach to Springer's theory, as developed by two of us in [BM] I, and then work in this alternative framework to establish this theorem.

In [BM] I, Springer's theory of Weyl group representations was derived from the Beilinson–Bernstein–Deligne–Gabber direct sum decomposition theorem [BBD], applied to Springer's resolution $\pi: T^*X \to \mathcal{N}$. The direct sum decomposition theorem gives that the sheaf $\mathbf{A}^\bullet := R\pi_* \mathbb{C}_{T^*X}$, considered as an object in the bounded derived category $\mathcal{D}^b(\mathcal{N})$ decomposes into direct summands $\mathbf{A}^\bullet_{(x,\varphi)}$, where x runs over a set of representatives of all nilpotent orbits, φ runs over the irreducible characters of $\pi_1(\mathcal{O}_x)$, and

$$(*) \qquad\qquad \mathbf{A}^\bullet_{(x,\varphi)} := Rj^x_* \, IC^\bullet(L_\varphi) \otimes V_{(x,\varphi)}$$

in the notion of [BM] I. We recall that $V_{(x,\varphi)}$ is a certain vector space, and the formula states that $\mathbf{A}^\bullet_{(x,\varphi)}$ is a sum of $\dim V_{(x,\varphi)}$ copies of the direct image under the inclusion map $j^x : \mathcal{O}_x \hookrightarrow \mathcal{N}$ of the intersection homology sheaf $IC^\bullet(L_\varphi)$ with coefficients in the local system L_φ of monodromy φ. To simplify notation we shall drop φ if $\varphi = 1$ is trivial, so write $\mathbf{A}^\bullet_{(x,1)} = \mathbf{A}^\bullet_x$ etc. Now application of the hypercohomology functor \mathbb{H}^i provides an isomorphism

$$\beta : H^i(X) \xrightarrow{\sim} \mathbb{H}^i(\mathcal{N}, \mathbf{A}^\bullet) \ ,$$

and the direct sum decomposition

$$\mathsf{H}^i(\mathcal{N},\mathbf{A}^\bullet) \cong \mathsf{H}^i\left[\mathcal{N}, \bigoplus_{(x,\varphi)} \mathbf{A}^\bullet_{(x,\varphi)}\right] \cong \bigoplus_{(x,\varphi)} \mathsf{H}^i(\mathcal{N},\mathbf{A}^\bullet_{(x,\varphi)})$$

provides an inclusion as a direct summand

$$\gamma : \mathsf{H}^i(\mathcal{N},\mathbf{A}^\bullet_x) \hookrightarrow \mathsf{H}^i(\mathcal{N},\mathbf{A}^\bullet)$$

in each degree i, for each nilpotent orbit \mathcal{O}_x.

Proposition: *There exists an isomorphism α which makes the following diagram commutative:*

$$
\begin{array}{ccc}
H^{c\ell}_{4n-2d}\overline{(\pi^{-1}\mathcal{O}_u)} & \xrightarrow{\ \delta\ } & H^{2d_u}(X) \\[2mm]
\cong \Big\downarrow \alpha & & \cong \Big\downarrow \beta \\[2mm]
\mathsf{H}^{2d_u}(\mathcal{N},\mathbf{A}^\bullet_u) & \xrightarrow{\ \gamma\ } & \mathsf{H}^{2d_u}(\mathcal{N},\mathbf{A}^\bullet)
\end{array}
$$

Clearly, this implies part a) of the theorem. We shall next explain in 1.10, how also parts b) and c) about the W–action will follow. Then the construction of the isomorphism α will be given in 1.11, 1.12, to complete the proof of the theorem.

1.10 The Weyl group action.

Let us now take into account also Borel's W action on $H^*(X)$ (1.7) resp. Lusztig's [Lu] W–action on \mathbf{A}^\bullet, and recall that these make the above isomorphism β W–equivariant (cf. [BM1], section 6, [Sp]). Furthermore, we conclude from the main theorem of [BM1], that the direct summands $\mathbf{A}^\bullet_{(x,\varphi)}$ are stable under this action, and that the W action on each $\mathbf{A}^\bullet_{(x,\varphi)}$ is given by a linear representation $\rho_{(x,\varphi)}$ of W on the vector space $V_{(x,\varphi)}$, which is *irreducible* and identifies with Springer's representation up to a sign character. Application of the hypercohomology functor H^i to formula 1.9 (*) gives

$$\mathsf{H}^i(\mathcal{N},\mathbf{A}^\bullet_{(x,\varphi)}) \cong IH^{i-2d_x}(\mathcal{O}_x,L_\varphi) \otimes V_{(x,\varphi)},$$

which describes the hypercohomology of each direct summand $A^{\bullet}_{(x,\varphi)}$ as a W–module, which is isotypical of type $\rho_{(x,\varphi)}$, with multiplicity given by the *intersection homology group* of $\overline{\mathcal{O}_x}$, with coefficients in L_{φ}, in the appropriate degree (notations as in [BM1]). Taking $x = u$, $\varphi = 1$, $i = 2d_u$ this intersection homology group becomes $IH^0(\overline{\mathcal{O}_u})$, which is spanned by a single canonical class $[\mathcal{O}_u]$, so that we find

$$\mathbb{H}^{2d_u}(\mathcal{N}, A^{\bullet}_u) \cong IH^0(\overline{\mathcal{O}_u}) \otimes V_u \cong V_u .$$

Or in other words, this says that the map γ in proposition 1.9 maps $\mathbb{H}^{2d_u}(\mathcal{N}, A^{\bullet}_u)$ onto a single copy of Springer's representation ρ_u. This shows that the proposition will also imply parts b) and c) of our theorem.

1.11 Reduction to a crucial lemma.

We now turn to the construction of an isomorphism α as announced in the proposition (1.9). As a preliminary, we shall replace our characteristic class map δ by a similar but more convenient map δ', using the following commutative diagram

$$
\begin{array}{ccc}
\mathbb{H}^{2d_u}(T^*X, T^*X - \overline{\pi^{-1}\mathcal{O}_u}) & \xrightarrow{\ \delta'\ } & \mathbb{H}^{2d_u}(T^*X) \\
\cong \downarrow \epsilon & & \cong \downarrow \sigma^* \\
H^{c\ell}_{4n-2d_u}(\overline{\pi^{-1}\mathcal{O}_u}) & \xrightarrow{\ \delta\ } & H^{2d_u}(X)
\end{array}
$$

which will allow us to work entirely in cohomology of T^*X. Here the map δ' is defined just by functoriality of cohomology on pairs of topological spaces, and ϵ is Lefschetz duality, which is an isomorphism since T^*X is nonsingular. In view of this diagram, proposition 1.9 is equivalent to the following:

Proposition: *There is an isomorphism α' making this diagram commutative*

$$H^{2d_u}(T^*X, T^*X - \overline{\pi^{-1}\mathcal{O}_u}) \xrightarrow{\ \delta'\ } H^{2d_u}(T^*X)$$

$$\cong \Big\downarrow \alpha' \qquad\qquad \cong \Big\downarrow \beta' = \beta\sigma^*$$

$$\mathbb{H}^{2d_u}(\mathcal{N}, A_u^\bullet) \xrightarrow{\qquad \gamma \qquad} \mathbb{H}^{2d_u}(\mathcal{N}, A^\bullet)$$

We prefer to prove this equivalent version. The construction of α' will be accomplished by a discussion of the following diagram, where we denote $\partial\mathcal{O}_u = \overline{\mathcal{O}_u} \backslash \mathcal{O}_u$ the topological boundary of \mathcal{O}_u, and $d = d_u$, to simplify notation.

In this diagram, all horizontal maps, and also the top triangles, are induced by inclusions of pairs of topological spaces. The middle row of vertical isomorphisms are examples of the sheaf–theoretic isomorphism

$$(*) \qquad\qquad \mathbb{H}^i(\pi^{-1}A, \pi^{-1}B; S^\cdot) \xrightarrow{\ \sim\ } \mathbb{H}^i(A, B; R\pi_* S^\bullet)$$

which holds for any map π of a space with a complex of sheaves S^\bullet to a space with a pair of subspaces $A \supset B$. Finally, the bottom vertical arrows are induced by coefficient inclusions $A_u^\bullet \to A^\bullet$; they are inclusions, since A_u^\bullet is a direct summand of A^\bullet. Each of the small squares and triangles in the diagram clearly commutes, so *the diagram is commutative.*

Lemma. *The maps ξ, η, ζ, ω in this diagram are isomorphisms.*

Given the lemma, we may define the desired isomorphism α' by tracing around the outside edge of the diagram $(\alpha' = \zeta\eta^{-1}\xi^{-1}v\omega)$, and then the commutativity of the whole diagram gives $\gamma\alpha' = \beta'\delta'$, which establishes our proposition.

1.12 Completion of the proof of theorem 1.8.

It is left to prove the lemma. To prove that ξ is an isomorphism, it suffices to check that for all direct summands $A^{\bullet}_{(x,\varphi)}$ other than A^{\bullet}_u, the group $H^{2d}(N\backslash\partial O_u, N\backslash\overline{O_u}; A^{\bullet}_{(x,\varphi)})$ vanishes. We may interpret this group as the global sections $\Gamma L_{(u,(x,\varphi))}$ of the local system $L_{(u,(x,\varphi))}$ on O_u whose fiber at $u \in O_u$ is given by $H^{4n-2d}(N, N\backslash\{u\}; A^{\bullet}_{(x,\varphi)})$. Now if $O_x \neq O_u$, then this fibre is zero by a dimension count, using the support conditions for intersection homology (axioms (AX2) in [GM2]). If $O_x = O_u$, but $\varphi \neq 1$, then the local system has no invariants. It follows that indeed

$$H^{2d}(N\backslash\partial O_u, N\backslash\overline{O_u}; A^{\bullet}_{(x,\varphi)}) = 0 \ , \ \text{ whenever } (x,\varphi) \neq (u,1) \ ,$$

and hence ξ is an isomorphism.

To see that η is an isomorphism, look at the case $(x,\varphi) = (u,1)$ of the previous discussion, and interpret the target of η as $\Gamma L = H^0(O_u, L)$ with $L = L_{(u,(u,1))}$, notation as above. Similarly, the source of η may be interpreted as $IH^0(\overline{O_u}, L)$, so we have a commutative diagram of canonical maps

$$\begin{array}{ccc} H^{2d}(N\backslash\partial O_u, N\backslash\overline{O_u}; A^{\bullet}_u) & \xrightarrow{\ \eta\ } & H^{2d}(N, N\backslash\overline{O_u}; A^{\bullet}_u) \\ \| & & \| \\ H^0(O_u, L) & \xleftarrow{\hspace{3cm}} & IH^0(O_u, L) \ . \end{array}$$

Now it is a general fact in intersection homology theory, that the bottom canonical map is an isomorphism: In the language of [GM1], codimension zero cycles are always allowable.

Finally, the map η is an isomorphism since A_u^\bullet is supported on $\overline{\mathcal{O}_u}$. And the map ω is an isomorphism because its source and target are both vector spaces with basis the irreducible components of $\overline{\pi^{-1}\mathcal{O}_u}$, and ω identifies them via these bases. This completes the proof of the lemma, and hence of the theorem. Q.e.d.

1.13 Comparison with Springer's original construction.

We have seen in theorem 1.8 that our characteristic class map $\delta : \mathcal{K} \to Q(\mathcal{K})$ embeds $H_{4n-2d_u}^{\mathcal{d}}(\overline{\pi^{-1}\mathcal{O}_u})$ into $H^{2d_u}(X)$ as a subspace invariant for Borel's W–action on $H^*(X)$, and we even know already that the W–action on this subspace is equivalent to ρ_u, that is to the Springer representation of W on $H^{2d_u}(\pi^{-1}u)^{G_u}$ (cf. 1.8). It remains to show how this equivalence can be realized by some geometrically defined map between the two representation spaces. This is the purpose of the theorem below, which complements theorem 1.8.

We shall use here the construction of the W–action on $H^*(\pi^{-1}u)$ as in [BM] 1, due to Lusztig [Lu], and which differs from Springer's originally defined representations by a multiplication with the sign character (cf. loc. cit., and also [Ho], Appendix, [Sp], §2 for more details). We shall actually prefer to work with the contragredient representation on the top homology space $H_{2d_u}(\pi^{-1}u)$, dual to $H^{2d_u}(\pi^{-1}u)$. This space, and similarly $H_{4n-2d_u}(\overline{\pi^{-1}\mathcal{O}_u})$, are equipped with canonical bases, given by the (canonical classes $[C]$ resp. $[\mathcal{K}]$ of) the irreducible components C resp. \mathcal{K} of $\pi^{-1}u$ resp. $\pi^{-1}\mathcal{O}_u$. So the linear maps linking these vector–spaces as explained below can be explicitly described by referring to these bases. Note that G_u acts via the finite group $C(u) := G_u/G_u^0$ of its connected components on (co)homology, since G_u^0 acts trivially. Since $C(u)$ acts on $H_{2d_u}(\pi^{-1}u)$ by permuting the canonical basis, we get also a canonical basis of the

subspace of G_u–invariants, corresponding to the $C(u)$–orbits of components C. We even get a canonical projection p of $H_{2d_u}(\pi^{-1}u)$ onto $H_{2d_u}(\pi^{-1}u)^{G_u}$ defined by

$$(1) \qquad\qquad p[C] = \sum_{a \in C(u)} [aC] = n_C([C_1]+...+[C_r])$$

where $C_1,...,C_r$ are the *different* $C(u)$–conjugates of C, and n_C is the number of $a \in C(u)$ fixing C. Starting from \mathcal{K}, an irreducible component of $\overline{\pi^{-1}\mathcal{O}_u}$, we recall from [BB] III, B.2 that

$$\mathcal{K} \cap \pi^{-1}u = G_u C = C_1 \cup ... \cup C_r$$

is the union of one full $C(u)$–orbit of components of $\pi^{-1}u$ (notation as before). We conclude that the Gysin homomorphism

$$j : H^{\mathcal{d}}_{4n-2d_u}(\overline{\pi^{-1}\mathcal{O}_u}) \longrightarrow H_{2d_u}(\pi^{-1}u)$$

is given by

$$(2) \qquad\qquad j[\mathcal{K}] = [G_u C] = [C_1] +...+ [C_r] .$$

In particular, it is a linear isomorphism onto the G_u–invariants of the target space. We also recall from loc.cit. that the G–saturation GC, that is the union of all $gC(g \in G)$, is an irreducible component of $\pi^{-1}\mathcal{O}_u$, so its closure is an orbital cone bundle $\overline{GC} = \mathcal{K}$, and we have a linear map q in the reverse direction of j by

$$(3) \qquad\qquad q[C] = n_C[\overline{GC}] ,$$

where the scale factor n_C is added to make the triangle p, j, q commute. Let us summarize this discussion as follows.

Proposition: *We have a commutative diagram of linear maps*

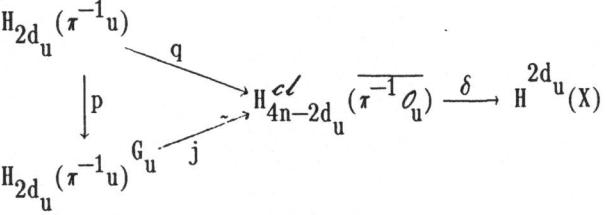

described in (1), (2), (3) *resp.* 1.8 *above.*

1.14 Theorem: *The maps in the above diagram are W−equivariant.*

In particular, the Weyl group representation on characteristic classes of orbital cone bundles *is equivalent to Springer's representation* ρ_u *via the Gysin map* j.

Proof. Note that the "inclusion" δ is W−equivariant by definition, and the projection p is W−equivariant since the actions of G_u and W commute [S1], [BM1]. It is therefore sufficient to prove that the linear isomorphism j is W−equivariant.

To do this, chose a transversal slice A in \mathfrak{g} to \mathcal{O}_u at u, and denote $D = A \cap \mathcal{N}$ its intersection with the nilpotent cone. (For instance, A may be chosen as an affine subspace of codimension $2d_u = \dim \mathcal{O}_u$ in \mathfrak{g} meeting the tangent space of \mathcal{O}_u only in the point u). Consider the following commutative diagram:

$$
\begin{array}{ccccc}
H_{2d_u}(\pi^{-1}u) & \longleftarrow & H^{cl}_{4n-2d_u}(\pi^{-1}\mathcal{O}_u) & \overset{\simeq}{\longleftarrow} & H^{cl}_{4n-2d_u}(\overline{\pi^{-1}\mathcal{O}_u}) \\
\uparrow{\simeq} & & \uparrow{\simeq} & & \uparrow{\simeq} \\
H^{2d_u}(\pi^{-1}D, \pi^{-1}D - \pi^{-1}u) & \leftarrow & H^{2d_u}(T^*X - \pi^{-1}\partial\mathcal{O}_u, T^*X - \pi^{-1}\mathcal{O}_0) & \overset{\omega}{\longleftarrow} & H^{2d_u}(T^*X, T^*X - \overline{\pi^{-1}\mathcal{O}_u}) \\
\downarrow{\simeq} & & v\downarrow{\simeq} & & \vdots \\
H^{2d_u}(D, D\backslash\{u\}, A^\bullet) & \leftarrow & H^{2d_u}(\mathcal{N} - \partial\mathcal{O}_u, \mathcal{N} - \overline{\mathcal{O}_u}, A^\bullet) & & \downarrow
\end{array}
$$

Our map j occurs in this diagram as the composition of the two Gysin

homomorphisms in the first row. The other horizontal arrows come from inclusions of pairs of topological spaces. The upper row of vertical arrows are Lefschetz duality isomorphisms, while the lower ones are again examples of the isomorphism 1.11 (*). Let us point out that the dotted arrow, analogous to v, would generally not be an isomorphism, because $\overline{\pi^{-1}\mathcal{O}_u}$ will be strictly smaller than $\pi^{-1}\mathcal{O}_u$ in general and is therefore <u>not</u> a π-preimage. This is the reason why we are inserting the middle column in the diagram.

Now observe that Springer's W-representation on $H_{2d_u}(\pi^{-1}u)$ is obtained from Lusztig's W action on \mathbf{A}^\bullet by transport of structure up the left column of this diagram. On the other hand, our W-action on $H_{4n-2d_u}(\overline{\pi^{-1}\mathcal{O}_u})$ is induced from Borel's on $H^{2d_u}(X)$, which is again obtained from the same W-action on \mathbf{A}^\bullet by transport of structure up the right column in the big diagram of 1.11. This shows that j is in fact W-equivariant.

1.15 Hotta's transformation formulas.

We may now describe quite explicitly in terms of integer matrices, how the characteristic classes $Q(\mathcal{K}_1),...,Q(\mathcal{K}_r)$ transform under the action of W (notations as in 1.8).

Theorem (cf. [Ho], Theorem 1 [Ho2]): *For each simple reflection s in W and each $i = 1,...,r$, we have either*

$$sQ(\mathcal{K}_i) = -Q(\mathcal{K}_i) ,$$

or else

$$sQ(\mathcal{K}_i) = Q(\mathcal{K}_i) + \sum_j n_{ij}^s \, Q(\mathcal{K}_j)$$

for certain non—negative integer coefficients n_{ij}^s, *which are zero unless* \mathcal{K}_j *intersects* \mathcal{K}_i *in codimension* 1.

Moreover, the coefficients n_{ij}^s can be described more precisely in geometrical terms, see Hotta's formula [Ho],1.5, definition 2 (cf. also 4.13 of the present book).

The fact that our classes $Q(\mathcal{K}_1),...,Q(\mathcal{K}_r)$ satisfy Hotta's transformation formulas may now be viewed as a corollary of theorem 1.14. However, another proof of this theorem (which does not assume Hotta's work), will also follow from chapter 4 of our present book (see 4.13).

§2. Generalities on equivariant K—theory

For the convenience of those readers not familiar with equivariant K—theory, we have collected here in some detail the general facts needed from this theory as prerequisites for subsequent chapters. In the present chaper, G may be an arbitrary linear algebraic group over k.

2.1 Algebraic notion of fibre bundle [Sr2]. [We].

By a G—*variety*, we mean an algebraic variety Y over k, equipped with an algebraic action of G on Y. A G—*morphism* $\varphi : Y \to X$ is a map of a G—variety Y into a G—variety X, which is a morphism of algebraic varieties, and respects the G—action. A surjective G—morphism is called a *principal* G—*fibration*, if G acts simply transitively on each fibre. The projection of X × G onto the left factor, with G acting only on the right factor, provides an example, refered to as *trivial*. A *principal fibre bundle* with structure group G, base X, and total space Y, is then defined as a principal G fibration $\varphi : Y \to X$, which is *locally trivial*, meaning that each point in X has a neighborhood \mathcal{U} such that $\varphi^{-1}\mathcal{U} \to \mathcal{U}$ is trivial (up to isomorphism). Note that, so far, the definitions are only a word by word translation from a topological or analytical context. In the algebraic context, however, there is a subtle point to be clarified here: The "local triviality" in the definition may refer to either the Zariski or the étale topology, that is \mathcal{U} above is a (Zariski–) open affine neighborhood in the first case, resp. any étale covering of such in the second case. For example, the principal fibration of G by an algebraic subgroup H need not be a bundle in the first, narrow sense (used by Weil [We]), but it is always a bundle in the second, wider sense (introduced by Serre [Sr2]). However, it fortunately turns out, for a lot of groups G, called "special" in [Sr2], that this subtle difference does not matter at all, that is to say local triviality in the weak ("étale") sense implies local triviality in the strong ("Zariski") sense for principal

fibrations with this structure group. "Special" groups in this sense include all connected solvable linear groups [Ro], and also $G = GL_n$ [Sr2]. Now let F be any G variety, and Y a principal G bundle over X in either sense. Then the *associated fibre bundle*

$$Y \times^G F \to X$$

with fibre F is defined as usual (see e.g. [Sr2]), and is locally trivial in the corresponding sense. So again, for the "special" groups above, the two notions of local triviality coincide for the associated fibre bundle as well.

For example, taking $F = k^n$ and Y a principal GL_n bundle on X, we get $Y \times^{GL_n} F$, an (algebraic) *vector bundle* over X. Since each vector bundle is obtained in this fashion (up to isomorphism), we conclude from the preceding remarks that *a vector bundle is locally trivial with respect to Zariski topology, if and only if it is locally trivial with respect to étale topology.* As a consequence of this discussion, we will not have to care anymore about the difference in this paper.

2.2 Equivariant vector bundles and definition of $K_G(X)$ [SGA6], [A1].

Now we assume that G acts on both the base X and the total space Y of a vector bundle, and that the bundle map $\varphi : Y \to X$ is a G morphism (2.1). Moreover, we assume that G preserves the linear structure, in the sense that each group element $g \in G$ maps the fibre $Y_x = \varphi^{-1}(x)$ (at $x \in X$) *linearly* into Y_{gx}. Then Y, equipped with this additional structure, is called a $G-$*equivariant vector bundle*, or just a G *vector bundle*. *Morphisms* of G vector bundles are defined as vector bundle homomorphisms which are simultaneously G morphisms.

We denote by $K_G(X)_{\mathbb{Z}}$ the Grothendieck group of the category of G vector bundles on X. We write

$$K_G(X) = K_G(X)_{\mathbb{Z}} \otimes_{\mathbb{Z}} k \ ,$$

according to our general conventions on coefficients (cf. §1). The formation of direct sums resp. tensor products on G vector bundles induces the structure of a commutative *ring* on $K_G(X)$. The *rank* of a vector bundle gives rise to an *augmentation* homomorphism $\epsilon : K_G(X) \rightarrow k$ of this ring, and the formation of exterior powers of G–vector bundles defines the so–called λ–operations λ^i : $K_G(X) \rightarrow K_G(X)$; this equips $K_G(X)$ with the structure of an "*augmented* λ–*ring*" in the sense of Grothendieck, as defined and studied from an axiomatic point of view in [SGA6], exposé V, or also in the first chapters of [Kn], [FL]. Our terminology here is completely analogous to Atiyah's, who considers topological vector bundles equivariant under a finite or compact group G in [A1], §§1.6, 2.3.

Remarks. 1) If G = 1 is the trivial group, then $K_G(X)$ reduces to the ring K(X) of "ordinary" K–theory (as considered already in §1), see e.g. [Ma],[FL], or [A1].

2) If X is a single point, then a G–vector bundle on X is just a G–*module*, that is a finite–dimensional linear representation of G on a k–vector space. Hence in this case, $K_G(X)$ is nothing else but the *representation ring* R(G) (cf. [A1]).

So equivariant K–theory is a common generalization of these two important extreme cases.

2.3 Equivariant homogeneous vector bundles.

As an important non–trivial example, let us consider equivariant vector bundles on a *homogeneous space*, say X = G/B, with *isotropy group* B any closed subgroup of G. Starting from an arbitrary B–module F (of finite dimension over k), let us form the associated fibre bundle $Y = G \times^B F$, which is a *vector bundle* on X (cf. the remark on local triviality in 2.1). We make it into a G–*equivariant* vector bundle, by making G act on G by left multiplication (whereas for the

bundle construction, B acted on G by right multiplication).

Proposition: *The construction of associated fibre bundles* $F \to G \times^B F$ *as explained above induces an isomorphism of augmented* λ–*rings:*

$$R(B) \xrightarrow{\sim} K_G(X) .$$

Obviously, the construction induces a λ–ring homomorphism and preserves the augmentation. It is not difficult to see that every G–vector bundle on X has the form $G \times^B F$ up to isomorphism (cf. e.g. [Se1], p. 130), which means that the homomorphism is *surjective*. An *inverse* homomorphism will be provided by restriction (to the isotropy group and the base point of X), cf. 2.9, below.

2.4 Functoriality in the group G.

If $\phi : B \to G$ is a morphism of algebraic groups, then our G–variety X becomes also a B–variety, and each G–equivariant vector bundle on X is also B–equivariant. This provides a functorial homomorphism

$$\phi^* : K_G(X) \longrightarrow K_B(X) ,$$

which we refer to as *restriction* from G to B. This is a homomorphism of *augmented* λ–*rings*.

If for instance $B = 1$, then this is a canonical homomorphism $K_G(X) \to K(X)$, refered to as the *forgetful* homomorphism, since it is given by "forgetting" the G–action on a G–vector bundle. In case $X = \{x\}$ is a single point for example, this forgetful homomorphism $K_G(X) \to K(X) \cong k$ is just the augmentation homomorphism $\epsilon : R(G) \to k$ (cf. 2.2).

2.5 Functoriality in the space X.

Let X, X' be two G–varieties, and $f : X \to X'$ a G–morphism. Then for each G–vector bundle E over X', the pull–back $f^* E$ to a vector bundle over X

is again G—equivariant, and this induces a homomorphism

$$f_G^* : K_G(X') \longrightarrow K_G(X) \, ,$$

which also preserves the structure of augmented λ—rings. In the case $G = \{1\}$
this is the so—called *Gysin homomorphism* $f^* : K(X') \rightarrow K(X)$ in ordinary
K—theory (cf. [A1], and [Fu], Example 15.1.8, or [SGA6], III.4.1, IV.2.7). The
G—equivariant analogue f_G^* *induces* the ordinary Gysin homomorphism, in the
sense that we have a commutative diagram

$$
\begin{CD}
K_G(X') @>{f_G^*}>> K_G(X) \\
@VVV @VVV \\
K(X') @>{f^*}>> K(X)
\end{CD}
$$

of "forgetful" and Gysin homomorphisms.

2.6 The sheaf theoretic point of view.

It will be convenient for us to work with G—*equivariant coherent sheaves* of
\mathcal{O}_X modules on our (irreducible) G—variety X, as we did in [BB] III. Let
$K_G(X)_0$ denote the Grothendieck group of the category of all such sheaves. (This
notational covention follows Fulton [Fu], p. 281). Since we can identify a vector
bundle over X with the locally free \mathcal{O}_X module of its germs of sections, and since
G—equivariance of the vector bundle means the same as G—equivariance of the
corresponding locally free \mathcal{O}_X module, we obtain a canonical homomorphism
$K_G(X) \rightarrow K_G(X)_0$. In the cases of interest for our present paper, this turns out to
be an *isomorphism* by the following proposition.

Proposition: *If* X *is a smooth* G—*variety which* (*) *admits a* G *equivariant ample
line bundle, then* $K_G(X) \cong K_G(X)_0$ *by the canonical homomorphism.*

Remarks. 1) Note that the existence of an ample line bundle means exactly that the variety under consideration is *quasi–projective*. Therefore, in the case $G = 1$, hypothesis (*) means X quasi–projective, and so the proposition coincides with Borel–Serre's Théorème 2 in [BS] for the non–equivariant case.

2) For example, the hypothesis (*) of this proposition is always satisfied for X *projective* and G *semi–simple*. In fact, according to Mumford [Mu], Chapter 1.3, Corollary 1.6, any normal projective G–variety admits a G–equivariant ample line bundle. More generally, it easily follows that Mumford's result extends to any G–variety X for which there exists a G–equivariant affine morphism $X \to Y$ to a normal projective G–variety. Hence the proposition implies as a special case the following criterion, which will suffice for our purposes in this paper:

Corollary: *Let* E *be the total space of a* G*–vector bundle over a smooth projective* G*–variety. Then* $K_G(E) \cong K_G(E)_o$, *if* G *is semi–simple.*

3) In the sequel, we shall always assume that X satisfies the assumptions of proposition 2.6, and we shall use this proposition to *identify* $K_G(X)$ with $K_G(X)_o$.

2.7 Existence of equivariant locally free resolutions.

For the *proof* of the equivariant Borel–Serre theorem (proposition 2.6 above), one may proceed by imitating the original proof in [BS], pp. 105–108. The only step in their argument which requires a new proof in our present, equivariant version is Lemma 10 in loc. cit., which assures that coherent sheaves admit locally free resolutions. So let us just state and prove the "equivariant version" of this lemma here.

Lemma: *Every* G*-equivariant coherent sheaf* \mathbf{F} *on* X *is a quotient of a* G*-equivariant locally free sheaf* \mathbf{E} *of finite type on* X.

Note that repeated application of this lemma will imply the following (with assumptions as in proposition 2.6):

Proposition: *Every* G*-equivariant coherent sheaf* \mathbf{F} *on* X *admits a finite resolution by* G*-equivariant locally free sheaves of finite type.*

Proof of the lemma: Let L denote a G-equivariant ample line bundle on X, which exists by assumption. Then the tensor product $\mathbf{F} \otimes L^n$ over \mathcal{O}_X of F with a sufficiently big number n of copies of L will be generated by global sections, cf. [Ha], p. 153. Then the natural morphism

$$\Gamma(X, \mathbf{F} \otimes L^n) \otimes_k \mathcal{O}_X \rightarrow \mathbf{F} \otimes L^n$$

is surjective, and we can even find a finite-dimensional k-subspace $E \subset \Gamma(X, \mathbf{F} \otimes L^n)$ such that the restriction map $E \otimes_k \mathcal{O}_X \rightarrow \mathbf{F} \otimes L^n$ is still surjective. So we have represented $\mathbf{F} \otimes L^n$ as a quotient of a locally free sheaf $E \otimes_k \mathcal{O}_X$, and it only remains to observe that this can be done equivariantly. Since F and L are G equivariant, so is $\mathbf{F} \otimes L^n$. Hence G acts on $\Gamma(X, \mathbf{F} \otimes L^n)$ algebraically, by locally finite linear endomorphisms. By enlarging the subspace E if necessary, we may assume that E is a G submodule. Now we have G actions on E and on \mathcal{O}_X, hence on the tensor product $E \otimes_k \mathcal{O}_X$. It is clear that the above surjective morphism $E \otimes_k \mathcal{O}_X \rightarrow \mathbf{F} \otimes L^n$ is G equivariant. Tensoring this morphism with the inverse L^{-n} of the invertible sheaf L^n, we obtain a surjective G-equivariant morphism of the locally free coherent sheaf

$$\mathbf{E} := (E \otimes_k \mathcal{O}_X) \otimes_{\mathcal{O}_X} L^{-n} = E' \otimes_k L^{-n} \text{ onto } \mathbf{F} \otimes L^n \otimes L^{-n} \cong \mathbf{F}.$$

Q.e.d.

Remark: For results of the same type, but with different hypothesis, we refer to Thomason [T1], Corollary 5.2 ("Seshadri's conjecture", see also [Sh]).

2.8 Remarks on Gysin homomorphisms in terms of coherent sheaves.

a) In order to define the ("equivariant" resp. "ordinary") Gysin homomorphisms (cf. 2.5)

$$f_G^* : K_G(X')_0 \longrightarrow K_G(X)_0 \text{ resp. } f^* : K(X') \longrightarrow K(X)_0$$

for a G–morphism $f : X \to X'$ in working with coherent \mathcal{O}_X–modules, one has to assume f to be a "perfect" morphism in the sense of [SGA6], Définition III.4.1, and IV.2.7, see also [Fu], Example 15.1.8. For example, if $f : X \to X'$ is a closed embedding, such that $f_* \mathcal{O}_X$ has a finite resolution by a complex E. of locally free \mathcal{O}_X'–modules, then f is *perfect*, and the Gysin homomorphism f^* (also called "homological intersection with X") is given by the formula

$$f^*[F] = \sum_i (-1)^i \, [\mathrm{Tor}_i^{X'} (\mathcal{O}_X, F)]$$

for any locally free \mathcal{O}_X' module. Here $\mathrm{Tor}_i^{X'} (\mathcal{O}_X, F)$ viewed as a \mathcal{O}_X'–module, is the i^{th} cohomology group of the complex E. $\otimes_{\mathcal{O}_X'}$ F (cf. loc. cit.). If the morphism f and the complex E. are G–*equivariant*, then the same formula holds for the equivariant Gysin homomorphism f_G^*. More generally, if f and E. are only B–equivariant, for a closed subgroup B ⊂ G, then by restricting first the group from G to B, then the space from X' to X, we obtain a composed restriction homomorphism $f_B^* : K_G(X') \to K_B(X)$ satisfying a similar formula.

b) The above asssumptions on F are satisfied e.g. for f the inclusion of a *complete intersection*, or for f a *regular embedding* (cf. [SGA6], Example III, 4.1.1, or [Fu], chapter 6.2, for a detailed treatment of this case on the level of Chow groups).

2.9 Equivariant K—theory on a vector bundle: Basic restriction techniques.

A crucial technique, frequently applied in this paper, will be the investigation of equivariant K—theory on the *total space* of a homogeneous vector bundle, via its restriction to a fibre, or alternatively to the zero section, and finally to a point. Since the morphisms involved are all *regular embeddings*, the preceding remarks (2.8) apply to the corresponding restriction homomorphisms (and these are the only three, very special, cases of 2.8 relevant for our present work).

Because of the fundamental significance of this restriction technique for the whole paper, let us explain it in more detail on a general level here. We assume that X is an irreducible G—variety satisfying the assumptions of proposition 2.6.

1) **Restriction to a point.** Consider the inclusion map $i : \{x\} \to X$ of a point x in X. Let $B = G_x$ be the isotropy group at x. Then we obtain a restriction homomorphism $i_B^* : K_G(X) \to R(B)$ by the remarks 2.8 a), b). For X a *homogeneous* G—*space*, that is $X = G/B$, this turns out (easily) to be an *inverse* of the homomorphism $R(B) \to K_G(X)$ coming from the formation of associated fibre bundles (see 2.3), and so i_B^* is an *isomorphism* $K_G(X) \cong R(B)$ in this case.

2) **Restriction to the zero section.** Now let E be the total space of a G—vector bundle over X, and let $\sigma : X \to E$ denote the zero section. Then the restriction homomorphism

$$\sigma^* : K(E) \to K(X)$$

known as "homological intersection with the zero section", has an equivariant analogue

$$\sigma_G^* : K_G(E) \to K_G(X)$$

by 2.8. It is well known to algebraic geometers that σ^* is always an *isomorphism* (see [SGA6], Exposé IX, Proposition 1.6, or on the level of Chow groups [Fu], Theorem 3.3). According to recent work of Thomason, this generalizes to the equivariant situation, that is to say σ_G^* *is an isomorphism* quite generally [T1],

Theorem 4.1. In the cases of interest for our present paper, we shall see more explicit reasons why this map is an isomorphism (see 3.7 and 4.3), independently of [T1].

3) **Restriction to a fibre.** With E a G-vector bundle on X as before, let $j : E_x \to E$ denote the inclusion of the fibre E_x at $x \in X$. Then we obtain a restriction homomorphism

$$j_B^* : K_G(E) \to K_B(E_x) \ , \ \text{where} \ B = G_x \ \text{as before.}$$

4) *Combining* examples (1), (2) (twice), and (3) into the following diagram of inclusions

$$
\begin{array}{ccc}
E & \xleftarrow{\ \sigma\ } & X \\
\Big\uparrow{\scriptstyle j} & & \Big\uparrow{\scriptstyle i} \\
E_x & \xleftarrow{\ \iota\ } & x
\end{array}
$$

we obtain a commutative diagram of restriction homomorphisms

$$
\begin{array}{ccc}
K_G(E) & \xrightarrow{\ \sigma_G^*\ } & K_G(X) \\
\Big\downarrow{\scriptstyle j_B^*} & & \Big\downarrow{\scriptstyle i_B^*} \\
K_B(E_x) & \xrightarrow{\ \iota_B^*\ } & R(B)
\end{array}
$$

Summarizing our present discussion, we may now state:

Proposition: *This is a commutative diagram of isomorphisms of λ-rings.*

2.10 Filtrations on $K_G(X)$.

a) *Grothendieck's γ-filtration* $... \supset K_G^a(X) \supset K_G^{a+1}(X) \supset ...$, is defined as follows (cf. [SGA6], V.3.10, or [A2], §12): Let $I_G = \mathrm{Ker}\ \epsilon$ denote the augmentation ideal, and γ^i (i = 0,1,2,...) the operators given by the formula

$$\gamma^i(Z) = \lambda^i(Z+i-1)$$

for all ring elements Z, where ϵ and the λ^i operations are defined in 2.2. Then $K_G^a(X)$ is the ideal generated by all monomials

$$\gamma^{i_1}(Z_1)\gamma^{i_2}(Z_2) \ \dots \ \gamma^{i_r}(Z_r) \ , \ \text{with } i_1 + i_2 + \dots i_r \geq a \ ,$$

$$\text{and } Z_1, \dots, Z_r \in I_G \ .$$

Then $K_G^0(X) = K_G(X), K_G^1(X) = I_G,$ and

$$K_G^a(X)\ K_G^b(X) \subset K_G^{a+b}(X)$$

for all $a, b \in \mathbb{N}$, i.e. $K_G(X)$ becomes a *filtered ring*. This filtration has the advantage of being obviously functorial (i.e. preserved by the various restriction or Gysin homomorphisms introduced above), since its definition refers only to the structure of $K_G(X)$ as an abstract "augmented λ-ring", which is indeed functorial. In particular, the various functorial homomorphisms considered induce also homomorphisms on the *associated graded rings*, for which we shall use a notation generalizing that introduced in 1.4, that is:

$$\mathrm{gr}\ K_G(X) = \bigoplus_{a \geq o} \mathrm{gr}_a K_G(X) = \bigoplus_{a \geq o} K_G^a(X)/K_G^{a+1}(X) \ .$$

b) **Topological filtration.** Alternatively, we may filter the ring $K_G(X)$ by co–dimension of supports. More precisely, let $K_G^a(X)'$ denote the Grothendieck group of G–equivariant coherent \mathcal{O}_X–modules with support of codimension \geq a. Then $... \supset K_G^a(X)' \supset K_G^{a+1}(X)' \supset ...$ is another descending filtration of $K_G(X)$ as a ring. Its comparison to the γ–filtration is a delicate problem in general. In the non–equivariant situation (special case G = 1), it is known that the γ–filtration coincides with the topological filtration, as we already mentioned in 1.3.

In general, the two filtrations will be obviously very different, see the next example.

2.11 Representation rings for example.

The γ–filtration on $R(G)$, for a *commutative* reductive group G, is given by the powers of the augmentation ideal $I_G = \text{Ker } \epsilon$:

$$R^a(G) := K_G^a(\text{point}) = I_G^a, \quad \text{for all } a \in \mathbb{N}.$$

(This follows from the fact that all irreducible representations are one–dimensional, cf. [A2], Corollary 12.4.) In general, we have only $I_G^a \subset R^a(G)$ from the definition, but the topology defined by the γ–filtration on $R(G)$ still coincides with the I_G–adic topology. The *completed representation ring* $\hat{R}(G)$ is defined as the completion of $R(G)$ with respect to this topology.

2.12 Application of equivariant K–theory to \mathcal{D}–modules.

In this section, Y is a smooth algebraic variety over k, equipped with an action $q : G \times Y \to Y$ of an algebraic group G. It will be convenient to introduce a weaker notion for "equivariant \mathcal{D}_Y–modules" than was considered in [BB] III, 2.2. First, the sheaf of algebras $\mathcal{D}_{G \times Y}$ on $G \times Y$, which is isomorphic to the external tensor product $\mathcal{D}_G \times \mathcal{D}_Y$ (notation [Gd]), contains $\mathcal{O}_G \times \mathcal{D}_Y$ as a subsheaf of algebras.

Definition: A *weakly G–equivariant* \mathcal{D}_Y–module is a \mathcal{D}_Y–module \mathcal{M} equipped with an isomorphism $\alpha : q^* \mathcal{M} \to p^* \mathcal{M}$ of $\mathcal{O}_G \times \mathcal{D}_Y$–modules, which satisfies a certain cocycle condition which ensures that α induces a *group action* of G on \mathcal{M} (see [Mu] for more details).

Notice that both $p^* \mathcal{M}$ and $q^* \mathcal{M}$ are $\mathcal{D}_{G \times Y}$–modules, hence are $\mathcal{O}_G \boxtimes \mathcal{D}_Y$–modules.

A good filtration $(\mathcal{M}_n)_{n\in\mathbb{Z}}$ of a weakly G–equivariant \mathcal{D}_Y–module \mathcal{M} is said to be G–equivariant if the isomorphism $\alpha : q^*\mathcal{M} \overset{\sim}{\to} p^*\mathcal{M}$ maps $q^*\mathcal{M}_n$ to $p^*\mathcal{M}_n$ (so that each \mathcal{M}_n is a H–equivariant sub \mathcal{O}_Y–module of \mathcal{M}).

A weakly G–equivariant \mathcal{D}_Y–module \mathcal{M}, which is coherent (as a \mathcal{D}_Y–module), always admits a G–equivariant good filtration. Indeed, let $(\mathcal{M}_n)_{n\in\mathbb{Z}}$ be any good filtration of \mathcal{M}; then the intersection \mathcal{M}'_n of \mathcal{M}_n and of $p_*q^*\mathcal{M}$, is \mathcal{O}_Y–coherent and $(\mathcal{M}'_n)_{n\in\mathbb{Z}}$ is a G–equivariant good filtration of \mathcal{M}.

Now for (\mathcal{M}_n) a G–equivariant good filtration of \mathcal{M}, the associated graded module $\mathrm{gr}(\mathcal{M}) = \bigoplus_n (\mathcal{M}_n/\mathcal{M}_{n-1})$ is a G–equivariant coherent \mathcal{O}_{T^*Y} module, and therefore determines a class $[\mathrm{gr}\,\mathcal{M}]$ in $K_G(T^*Y)$. The lemma below asserts that *this class does not depend on the choice of* (\mathcal{M}_n). Hence a weakly G–equivariant \mathcal{D}_Y–module \mathcal{M} determines a well–defined class in $K_G(T^*Y) \cong K_G(Y)$, which we call the *class* of \mathcal{M} *in* $K_G(Y)$.

Lemma: Let $(\mathcal{M}_n)_{n\in\mathbb{Z}}$ and $(\mathcal{M}'_n)_{n\in\mathbb{Z}}$ be two G–equivariant good filtrations of \mathcal{M}. Then the corresponding associated graded modules $\mathrm{gr}\,\mathcal{M}$ resp. $\mathrm{gr}'\,\mathcal{M}$ determine the same class in $K_G(T^*Y)$.

Proof: By changing the numbering of the second filtration (if necessary), we may assume without loss of generality that $\mathcal{M}_n \subset \mathcal{M}'_n$ for all n. It is known that a for suitable integer $d \geq 0$, we have $\mathcal{M}'_n \subset \mathcal{M}_{n+d}$ for all n. Our proof proceeds by induction on d. Assume first $d = 1$. Then we have, for all $n \in \mathbb{Z}$, exact sequences of G–equivariant coherent \mathcal{O}_Y–modules

$$0 \longrightarrow \mathcal{M}'_n/\mathcal{M}_n \longrightarrow \mathcal{M}_{n+1}/\mathcal{M}_n \longrightarrow \mathcal{M}_{n+1}/\mathcal{M}'_n \longrightarrow 0$$
$$0 \longrightarrow \mathcal{M}_{n+1}/\mathcal{M}'_n \longrightarrow \mathcal{M}'_{n+1}/\mathcal{M}'_n \longrightarrow \mathcal{M}'_{n+1}/\mathcal{M}_{n+1} \longrightarrow 0 \ .$$

Hence we get exact sequences of G–equivariant coherent graded \mathcal{O}_{T^*Y}–modules.

$$0 \to A \to \mathrm{gr}\, \mathcal{M} \to B \to 0$$

$$0 \to B \to \mathrm{gr}'\mathcal{M} \to A \to 0,$$

where we put

$$A := \bigoplus_n \mathcal{M}'_n/\mathcal{M}_n\,, \text{ and } B := \bigoplus_n \mathcal{M}_{n+1}/\mathcal{M}'_n\,.$$

So in the group $K_G(T^*Y)$, we get the equality

$$[\mathrm{gr}\,\mathcal{M}] = [A] + [B] = [\mathrm{gr}'\mathcal{M}]\,.$$

Now the induction step: Assume for some $d > 1$ that $\mathcal{M}_n \subset \mathcal{M}'_n \subset \mathcal{M}_{n+d}$ for all n. We consider the further filtration of \mathcal{M} given by $\mathcal{M}''_n := \mathcal{M}_n + \mathcal{M}'_{n-1}$. It is again G–equivariant. It is also a *good* filtration, because for n big enough, we have

$$\mathcal{M}''_n := \mathcal{M}_n + \mathcal{M}'_{n-1} = \mathcal{D}_Y(1)\,\mathcal{M}_{n-1} + \mathcal{D}_Y(1)\mathcal{M}'_{n-2} = \mathcal{D}_Y(1)\mathcal{M}''_{n-1}\,.$$

We observe that this new filtration satisfies $\mathcal{M}_n \subset \mathcal{M}''_n \subset \mathcal{M}_{n+d-1}$. Hence $[\mathrm{gr}\,\mathcal{M}] = [\mathrm{gr}\,\mathcal{M}'']$ by the induction hypothesis. On the other hand, $\mathcal{M}'_{n-1} \subset \mathcal{M}''_n \subset \mathcal{M}'_n$ for all n. So we know by the $d = 1$ case of the proof that also $[\mathrm{gr}\,\mathcal{M}'] = [\mathrm{gr}\,\mathcal{M}'']$. This completes the proof of the lemma.

§3. Equivariant K–theory of torus actions and formal characters.

In this paragraph, we consider a torus T, that is a commutative connected reductive group over k, and a linear action of T on a vector space E of finite dimension r over k. We shall assume that *all weights* of T in E *are positive* with respect to some linear partial ordering \geq. We assume that the semi–group of positive (integral) weights in finitely generated. For example, T might be the group of homotheties of E. In the applications in subsequent chapters, T will be the maximal torus in a semisimple group, E will be the nilradical of a Borel subalgebra.

3.1 The completed representation ring of a torus.

The characters $\chi : T \to G_m$ of our torus T form a free abelian group $X^*(T)$, and the representation ring of T is isomorphic to the group ring of this character group, that is

$$R(T)_{\mathbb{Z}} \cong \mathbb{Z}[X^*(T)] \text{ , and}$$
$$R(T) \cong k[X^*(T)] \cong \mathcal{O}(T) \ .$$

The differential of $\chi \in X^*(T)$ is a linear form $d\chi : t \to k$ on the Lie algebra t of T, called an *integral weight*. Let $\Lambda = \Lambda(T)$ denote the lattice in t^* of all integral weights. For each $\lambda \in \Lambda$, we define the formal power series

$$e^\lambda := \sum_{i \geq o} \frac{1}{i!} \lambda^i \ ,$$

considered as an element in $\hat{S}(t^*)$, the completion of the ring $S(t^*)$ of polynomial functions on t (with respect to the $t*S(t*)$–adic topology.) We define a homomorphism $\varphi : R(T) \to \hat{S}(t*)$ by putting

$$\varphi(\chi) := e^{d\chi}$$

for all $\chi \in X^*(T)$, and we observe (cf. e.g. [AH], proposition 4.3) that this extends to an *isomorphism* (notation 2.11)

$$\varphi : \hat{R}(T) \overset{\sim}{\to} \hat{S}(t^*) \ .$$

This isomorphism will allow us to interpret the elements of the completed representation ring $\hat{R}(T)$ as formal power series functions on the Lie algebra. Note that this is an isomorphism of *filtered rings*. Let us make this slightly more explicit. Given a power series $P \in \hat{S}(t^*)$, we shall use the notation $[P]^i$ for its degree i homogeneous term, that is the unique homogeneous polynomial of degree i on t that

$$P = \underset{i \geq 0}{\Sigma} \ [P]^i \ .$$

Recalling that the γ–filtration on $\hat{R}(T)$ is given by the I_T–adic filtration, where I_T is the augmentation ideal (2.11), we get the following

Lemma: *If an element* $Q \in \hat{R}(T)$ *corresponds to the power series* $P = \varphi(Q) \in \hat{S}(t^*)$ *then its degree with respect to the* γ *filtration is equal to the smallest number* a *such that* $[P]^a \neq 0$.

Remark. We shall then refer to $[P]^a$ as the *lowest order term* of the series, and denote it also

$$\mathrm{gr} \ P := [P]^a \ .$$

3.2 Formal characters of T modules

Let M be a T–module, that is to say a k–vector space equipped with a linear algebraic action of T. For each character $\chi \in X^*(T)$ with differential $d\chi = \lambda \in \Lambda$, we denote by M_χ or M_λ the corresponding "weight space"

$$M_\chi = M_\lambda = \{v \in M \mid tv = \chi(t)v, t \in T\} \ .$$

Since T is a reductive group, $M = \underset{\lambda \in \Lambda}{\otimes} M_\lambda$ is a direct sum of weight spaces. If the *weight multiplicities* $\dim M_\chi$ are all *finite*, then we call M *admissible* (or we

say that M admits a formal character), and we *define* the *formal character* of M as the formal sum

$$\text{ch}(M) := \sum_{\chi} (\dim M_\chi)[\chi] \,.$$

We note that this definition coincides with the one used e.g. in [Di], 7.5, and [Ja], 4.5. Let us call ch(M) *bounded* (by χ_0), if its nonzero coefficients occur only at characters $\chi \leq \chi_0$, for some $\chi_0 \in X(T)$. Then the multiplication in R(T) extends to a multiplication of bounded formal characters, defined formally by

$$(\sum_{\xi \leq \xi_0} a_\xi [\xi]) (\sum_{\zeta \leq \zeta_0} b_\zeta [\zeta]) = \sum_{\chi \leq \chi_0} c_\chi [\chi] \,,$$

where

$$c_\chi := \sum_{\xi \zeta = \chi} a_\xi b_\zeta \quad \text{for all } \chi \in X(T),$$

and $\chi_0 = \xi_0 \zeta_0$. This makes the group of all bounded formal characters into an extension ring of R(T), which is denoted $R_{-\infty}(T)$. This ring may be described as a power series field $R_{-\infty}(T) \cong k[[x_1,...,x_\ell]]$, where $x_1,...,x_\ell$ are the *negative* fundamental weights. (In particular, $R_{-\infty}(T)$ should not be confused with the completed representation ring $\hat{R}(T)$.) Note that $R_{-\infty}(T)$ contains the *field of fractions* of R(T).

3.3 Example.

We consider here a finite dimensional T–module E whose weights are all *positive* with respect to the fixed partial ordering (see the beginning of this chapter). Then *the symmetric algebra* M = S(E) *admits a formal character*, which is bounded, and (with respect to multiplication in $R_{-\infty}(T)$) given by the formula

(1) $$\text{ch}(S(E)) = \Delta(E)^{-1}.$$

where we define

(2) $$\Delta(E) := \prod_\lambda (1-e^\lambda)^{\dim E_\lambda} \,,$$

the product being extended over all weights of E. In fact, for the special case where

$E = E_\lambda$ is one–dimensional, one gets immediately the geometric series

$$ch(S(E_\lambda)) = 1 + e^\lambda + e^{2\lambda} + ... = (1-e^\lambda)^{-1}$$

and then the general case follows by repeated application of the formula

$$ch(S(E \oplus F)) = ch(S(E) \otimes S(F)) = ch(S(E)) \cdot ch(S(F)) ,$$

which holds for any E, F satisfying the above positivity assumption.

As a corollary, we note that

(3)
$$ch(\mathcal{O}(E)) = \Delta(E^*)^{-1} = \prod_\lambda (1-e^{-\lambda})^{-\dim E_\lambda} ,$$

product over all weights λ of E. This is because $\mathcal{O}(E) \simeq S(E^*)$, and the weights of the contragredient T–module E^* have opposite signs, that is $\dim E^*_\lambda = \dim E_{-\lambda}$ for all $\lambda \in \Lambda$.

3.4. T–equivariant modules with highest weight.

Let M be a finitely generated $S(E^*)$–module, which is equipped with a linear, locally finite T–action, such that

$$t(sm) = (ts)(tm) \text{ for all } t \in T , m \in M , s \in S(E^*) .$$

For short, we say that M is a T–*equivariant* finitely generated $S(E^*)$–module. It is easy to see that such M admits a formal character. Let us be slightly more specific.

Consider a cyclic $S(E^*)$ submodule $M' = S(E^*)v$ generated by some weight vector $0 \ne v \in M$, of weight λ, say. Then $M' = kv + M''$, where $M'' = E^*M'$ has only weights strictly smaller than λ, because of our positivity assumption on the weights of E. Therefore λ is called *the highest weight* of M', and v resp. M' is called a cyclic *highest weight vector* resp. (sub–)*module of highest weight* λ. Now let I be the ideal of elements in $S(E^*)$ annihilating v. Then obviously I is a T submodule, and

(*)
$$ch(M') = e^\lambda \, ch(S/I) .$$

Here and in the sequel, we sometimes write $S = S(E^*)$, for short.

Lemma: a) *A finitely generated* T*—equivariant* $S(E^*)$ *module* M *admits a composition series*

$$0 = M_0 \subset M_1 \subset M_2 \subset ... \subset M_\ell = M$$

of T*—equivariant* $S(E^*)$ *submodules* M_i, *such that the composition factors* M_i/M_{i-1} *are cyclic highest weight modules.*

b) *Then* M *admits the formal character*

$$ch(M) = \sum_{i=1}^{\ell} e^{\lambda_i} ch(S/I_i) ,$$

where λ_i *is the highest weight of* M_i/M_{i-1}, *and* I_i *the annihilator of the corresponding cyclic highest weight vector.*

c) *The composition series can be chosen in such a way, that* $I_1,...,I_\ell$ *are prime ideals.*

Proof: Choose $M_1 = M'$ as in the preceding discussion, then repeat the same discussion for M/M_1 etc. Since M is noetherian, this process will terminate after a finite number of steps, proving a). Now b) follows from (*). Assuming $M \neq 0$, let P be a minimal associated prime ideal of the $S(E^*)$ module M. Since T is connected and acts by automorphisms on M, it stabilizes P. By equivariance, T stabilizes also the submodule of M annihilated by P, which is therefore a T—equivariant submodule $N \neq 0$. By choosing the weight vector $0 \neq v \in M$ in the discussion preceding the lemma even in N, we can achieve there $I = P$ prime. Repetition of this procedure proves part c) of the lemma. Q.e.d.

3.5. Projective and free cyclic highest weight modules.

Proposition. *Each projective finitely generated* T*—equivariant* $S(E^*)$ *module* M

admits a finite composition series with all composition factors free cyclic highest weight modules.

Here the terms "free" resp. "projective" refer to the $S(E^*)$ module structure. The proposition follows by repeated application of the following lemma.

Lemma: *A maximal weight space M_λ generates a free highest weight submodule $N = S(E^*)M_\lambda \cong S(E^*) \otimes_k M_\lambda$, and the quotient M/N is again projective.*

Proof: Let $m = E^* S(E^*)$ denote the maximal ideal corresponding to the zero point $0 \in E$. Consider the canonical linear map of $\tilde{N} := S(E^*) \otimes M_\lambda$ into M, which maps \tilde{N} onto N. The weight space M_λ injects into $M/m\,M$, because the weights in $m\,M$ are all smaller than λ. Hence the map from $S(E^*) \otimes_k M_\lambda$ to M induces an injection on the modules tensored with $S(E^*)/m\,S(E^*)$. Because both modules are projective over $S(E^*)$, it follows that $N = S(E^*) \otimes_k M_\lambda$ is a submodule of M, and that the quotient is locally free at m. Hence M/N is locally free in a neighborhood $U \subset E$ of the zero point. But then the T–saturation is $TU = E$ (since we assume $\sum_{\lambda > 0} E_\lambda = E$), and since M/N is T–equivariant, it is locally free everywhere on E. This means that M/N is projective. Q.e.d.

Corollary: *The formal character of M is given by*

$$\mathrm{ch}(M) = \Delta(E^*)^{-1}(e^{\lambda_1} + ... + e^{\lambda_\ell}) \ ,$$

where $\lambda_1, ..., \lambda_\ell$ are the highest weights of the free cyclic composition factors mentioned in the proposition.

In fact, for a free cyclic highest weight module M_I of highest weight λ_i we clearly have

$$\mathrm{ch}(N_i) = e^{\lambda_i}\,\mathrm{ch}(S) = e^{\lambda_i}\,\Delta(E^*)^{-1} \ ,$$

by combining 3.4(*) and 3.3(1).

Remark: The proposition states that, in the framework of lemma 3.4, the case "M projective" means that the prime ideals $I_1,...,I_\ell$ there are all zero; hence the corollary is also a special case of 3.4b).

3.6. Formal characters of equivariant coherent sheaves.

Now let $\text{Mod}_T^{\text{coh.}}(E)$ denote the category of T–equivariant coherent sheaves of \mathcal{O}_E modules, and $\text{mod}_T^{\text{f.g.}}(S(E^*))$ the category of T–equivariant finitely generated $S(E^*)$ modules. Then the functor $F \longmapsto \Gamma(E,F) = M$ establishes an *equivalence* of these categories, because E is affine. Moreover, a locally free F corresponds to a projective M, and since each F admits a finite *locally free resolution* in $\text{Mod}_T^{\text{coh.}}(E)$ by 2.7, we have simultaneously that each M admits a *projective resolution* in $\text{Mod}_T^{\text{f.g.}}(S(E^*))$. In particular, the (isomorphic) Grothendieck K groups of these categories are generated (over \mathbb{Z}) by locally free resp. projective objects, and so it follows from the preceding proposition 3.5 that they are even *generated by the free cyclic highest weight modules*. By abuse of language, the formal character of $M = \Gamma(E,F)$ is also called the formal character of F, notation $\text{ch}(F) := \text{ch}(M)$. Now it follows from the preceding discussion and from corollary 3.5 that $\Delta(E^*)\text{ch}(F)$ is an integer linear combination of exponentials.

Corollary: *For an arbitrary* T*–equivariant coherent* \mathcal{O}_E *module* F *(resp. finitely generated* $S(E^*)$ *module* $M = \Gamma(E,F)$*) the formal character is of the form*

$$\text{ch}(F) = \text{ch}(M) = \Delta(E^*)^{-1} \sum_{\lambda \in \Lambda} a_\lambda e^\lambda \ ,$$

where $a_\lambda = a_\lambda(F)$ *are integer coefficients, only finitely many of them being nonzero.*

Obviously, these integers a_λ are uniquely determined by **F** resp. M. Following traditional terminologies (cf. [Ja]), we refer to a_λ as the (integer) *multiplicity* of the free cyclic highest weight module of highest weight λ in the module M, notation

$$a_\lambda(F) = a_\lambda(M) = (M:gr\ M(\lambda))\ .$$

In the sequel, we consider $ch(F)$ as an element of the fraction field $Fract(R(T))$ of the domain $R(T)$.

3.7. Restriction to the zero point.

Since our T—action on E is linear, it fixes the zero point, and so the inclusion $\iota : \{0\} \to E$ gives rise to a restriction homomorphism $\iota_T^* : K_T(E) \to R(T)$.

Proposition: a) ι_T^* *is an isomorphism of* λ—*rings* $K_T(E) \overset{\sim}{\to} R(T)$.

 b) *This isomorphism can be computed within* $Fract(R(T))$ *by the following formula:*

$$\iota_T^*[F] = \Delta(E^*)ch(F)$$

which holds for any T—*equivariant coherent sheaf* **F** *on* E.

Comments. 1) We note that a) may be viewed as a very special case of a theorem of Thomason [T1], Theorem 4.1 on arbitrary equivariant vector bundles E, cf. 2.9(2). But it is convenient to prove the whole proposition here more directly below.

2) The proposition holds even with coefficients in \mathbb{Z}.

Proof: a) Since ι_T^* preserves the λ—ring structure by functoriality, it suffices to prove that it is *bijective*. In fact, let us exhibit the inverse map. Starting from an arbitrary finite—dimensional T module F, we form the locally free coherent \mathcal{O}_E

module

(1) $$F = \mathcal{O}_E \otimes_k F \ ,$$

and make T act diagonally on it. Then $\psi : [F] \to [\mathbf{F}]$ defines a homomorphism of $R(T)$ into $K_T(E)$. It is clear that \mathbf{F} has fibre F at the zero point, so

(2) $$\iota_T^*[\mathbf{F}] = [F] \ .$$

In particular, ψ is *injective*. We may even conclude from (2) that ψ is an inverse of ι_T^*, provided that we know ψ is *surjective*.

To see this, it suffices to show that *the sheaves of the particularly nice form* (1) *are sufficient to generate the whole Grothendieck group* $K_T(E)$. Observe that each locally free \mathbf{F} of the form (1) gives rise to a *projective* module $M := \Gamma(E, \mathbf{F})$ of the form $M \cong S(E^*) \otimes_k F$, (under the equivalence discussed in 3.6), *and conversely.* But the free cyclic highest weight modules are of this form, and we have observed in 3.6, that even these suffice to generate the full Grothendieck group under consideration. A fortiori, the sheaves of the form (1) generate $K_T(E)$ as a group.

b) Since both sides of the formula claimed in b) are additive in $[\mathbf{F}]$, it suffices to verify the formula only on a nice set of generators of $K_T(E)$, for example, just for those sheaves \mathbf{F} of the form (1). Assuming \mathbf{F} of this form, we get

$$M := \Gamma(E, \mathbf{F}) = \Gamma(E, \mathcal{O}_E) \otimes_k F = S(E^*) \otimes_k F \ ,$$
$$\mathrm{ch}(\mathbf{F}) = \mathrm{ch}(M) = \mathrm{ch}(S(E^*) \otimes_k F) = \mathrm{ch}(S(E^*))\mathrm{ch}(F) \ ,$$

and so by 3.3

$$\Delta(E^*)\mathrm{ch}(\mathbf{F}) = \mathrm{ch}(F) \ .$$

On the other hand, as pointed out in (2), we have

$$\mathrm{ch}(F) = [F] = \iota^*[\mathbf{F}] \ ,$$

which completes the proof of our proposition. Q.e.d.

3.8. Computation of γ degree.

Proposition 3.7a) says, in particular, that the isomorphism

$\iota_T^* : K_T(E) \overset{\sim}{\to} R(T)$ preserves degrees with respect to the γ—filtrations, and then

formula b) of the proposition tells us how we may compute the γ—degree of F, as

an element of $K_T(E)$, from the formal character $ch(F)$, considered as a formal

power series on t (cf. 3.1):

Corollary: *For any T—equivariant coherent sheaf* F *on* E, *the* γ *degree of* [F] *in*
$K_T(E)$ *is equal to the degree of the lowest order term of the power series*
$\Delta(E^*)ch(F)$ *in* $\hat{S}(t^*)$.

Using our notation for lowest order term as introduced in 3.1, we may write

this statement as a formula:

$$\gamma\text{—deg}[F] = \deg \, gr(\Delta(E^*)ch(F)) \, .$$

We shall determine the right hand side more explicitly below (3.10). Before doing

this, let us first look more systematically at the lowest order term of $\Delta(E^*)ch(F)$.

3.9. Character polynomials.

Let F be a T—equivariant coherent sheaf on E, and let $a = \gamma\text{—deg}[F]$

denote its γ—degree. By the filtered isomorphisms $K_T(E) \overset{\sim}{\to} R(T)$ (3.7) and

$\hat{R}(T) \overset{\sim}{\to} \hat{S}(t^*)$ (3.1), we may identify the associated graded ring of $K_T(E)$,

$$gr \, K_T(E) \overset{\sim}{\to} S(t^*)$$

with the graded ring of polynomial functions on t, by identifying for each degree j

the vector spaces

$$K_T^j(E)/K_T^{j+1}(E) \overset{\sim}{\to} \hat{S}_j(t^*)/\hat{S}_{j+1}(t^*) \overset{\sim}{\cong} S^j(t^*) \, .$$

In this manner, the class [F] modulo $K_T^{a+1}(E)$ determined by the sheaf F is

identified with a *homogeneous polynomial* of *degree* a *on* t.

Definition: We call this polynomial the *character polynomial* of \mathbf{F}, and denote it by $q_{\mathbf{F}}^{T}$. For a module $M = \Gamma(E,F)$, we use the similar notation and terminology, calling $q_{M}^{T} = q_{\mathbf{F}}^{T}$ the character polynomial of M. Normally, we drop the superscript T, if the torus of reference is clear enough from the context.

The term "character polynomial" refers to the fact that this polynomial is computable from the formal character by means of formula 3.7b) as the lowest degree term of the series $\Delta(E^{*})\mathrm{ch}(F)$:

Corollary: $q_{\mathbf{F}} = q_{M} = \mathrm{gr}\ \Delta(E^{*})\mathrm{ch}(F)$.

Let us illustrate the use of this formula by a few immediate applications.

Proposition: $q_{M} = \frac{1}{a!} \underset{\lambda \in \Lambda}{\Sigma}\ a_{\lambda}(M)\lambda^{a}$, $\hspace{3cm}$ (*)

where the integer $a_{\lambda}(M)$ is the "multiplicity" of the free cyclic highest weight module of highest weight λ in M (notation 3.6). Moreover, the number $a = \gamma - \deg[M]$ may be computed from these integers $a_{\lambda}(M)$ as the smallest positive integer a for which the right hand side of () becomes a nonzero polynomial.*

Proof: This is now an immediate consequence of 3.5, by writing out the sum of exponentials given there as a power series, and taking the lowest order term. Q.e.d.

Example: Let us compute the character polynomial q_{M} for $M = k$ the trivial, one-dimensional module. In this case, $\mathrm{ch}(M) = 1$, and so q_{M} is the lowest degree term of

$$\Delta(E^{*})\mathrm{ch}(M) = \Delta(E^{*}) = \prod_{i=1}^{r}(1-e^{-\lambda_{i}}) = \prod_{i=1}^{r}(1-1+\lambda_{i}-\tfrac{1}{2}\lambda_{i}^{2}+...) = \prod_{i=1}^{r}(\lambda_{i}) + ...$$

where $\lambda_{1},...,\lambda_{r}$ are the weights of T in E, and dots are terms of higher degree. Hence q_{M} is equal to the product of all weights of T in E, and in particular $\deg q_{M} = r := \dim E$ is the vector space dimension of E in this case. Similarly,

one obtains for an *arbitrary* M of *finite vector–space dimension* the formula

$$q_M = \dim M \prod_{i=1}^{r} (\lambda_i) \ ,$$

$$\deg q_M = \dim E =: r \ .$$

3.10 Degree of character polynomial equals codimension of support.

Theorem: *Let* F *be a* T*–equivariant coherent sheaf on* E. *Then*

(1) $\deg q_F = \mathrm{codim}_E \ \mathrm{supp} \ F$.

Comments: Note that, by definition of the character polynomial q_F, its degree is also the γ degree of [F]. Let us also restate the theorem in terms of the corresponding T–equivariant module $M = \Gamma(E,F)$:

(2) $\deg q_M = r{-}d(M)$,

where $r := \dim E$, and $d(M)$ denotes Krull (or Gelfand–Kirillov) dimension. We refer to Joseph [J1], 2.4 (ii) for a similar result; the precise relationship is explained later in our present paper (4.8).

Our proof of this theorem, given in 3.12–3.13, proceeds by induction on $d(M)$. To make the induction argument work, however, we first have to sharpen the theorem by the following technical complement.

3.11 Positivity property of character polynomials.

Let us call a polynomial q of t^* *positive*, if it takes only positive values on all positive integral weights.

Complement of Theorem 3.10: *The polynomial* q_F *is positive.*

Our induction argument, which will prove the theorem and its complement simultaneously, is based on the following key lemma.

3.12 Division by a non–zero divisor.

Lemma: *Let* $M \neq 0$ *be a finitely generated* T*–equivariant* $S(E^*)$ *module, and let* $0 \neq f \in S(E^*)$ *be a weight vector of weight* $-\mu \neq 0$, *which acts as a nonzero divisor on* M. *Then we have*

(1)
$$q_{M/fM} = \mu q_M$$

and

(2)
$$d(M/fM) = d(M) - 1.$$

Proof: In fact, the assumptions on f imply that multiplication by f maps each weight space M_λ injectively into $M_{\lambda-\mu}$, so that

$$ch(fM) = ch(M)e^{-\mu},$$

hence

$$ch(M/fM) = ch(M) - ch(fM) = (1-e^{-\mu})ch(M).$$

Since $1-e^{-\mu} = \mu - \frac{1}{2}\mu^2 + ...$, it follows that the lowest order term of $\Delta(E^*)ch(M/fM)$ is obtained from that of $\Delta(E^*)ch(M)$ by multiplication with μ. This proves (1). The last equation (2) is a well known property of Krull dimension.

$$\text{Q.e.d.}$$

As a consequence of (1), we can conclude that q_M *will be positive if* $q_{M/fM}$ *is*.

3.13 Proof of theorem 3.10 and 3.11.

Let us proceed by induction on $d(M)$. The case $d(M) = 0$ is settled by example 3.9. So assume $d(M) > 0$. We choose prime ideals $I_1,...,I_\ell$ as in 3.4c), so that by 3.4b):

$$ch(M) = e^{\lambda_1}ch(S/I_1) + ... + e^{\lambda_\ell}ch(S/I_\ell).$$

By 3.9, q_M is the lowest order term of the series

$$ch(M)\Delta(E^*) = \sum_i e^{\lambda_i}ch(S/I_i)\Delta(E^*). \qquad (*)$$

Let $\mathcal{J} \subset \{1,...,\ell\}$ denote the set of indices i for which $\deg q_{S/I_i}$ assumes its minimum value, m say. Observing that multiplication by an exponential series e^{λ_i} does not affect the lowest order term of a power series, we see that the power series (*) has its degree m term equal to $\sum_{i \in \mathcal{J}} q_{S/I_i}$, and has only zero terms in smaller degree. So we may conclude from (*) that $\deg q_M = m$, and

$$q_M = \sum_{i \in \mathcal{J}} q_{S/I_i}, \qquad (**)$$

unless this sum vanishes. But suppose for a moment we knew already that theorem 3.10 and its complement hold for the modules $M = S/I_i$ involved in the sum. Then the polynomials q_{S/I_i} are all *positive*, and so their sum (**) obviously cannot vanish, but must be positive as well. Hence

$$\deg q_M = m := \min_{1 \le i \le \ell} \deg q_{S/I_i} .$$

On the other hand we have

$$r{-}d(M) = \min_{1 \le i \le \ell} (r{-}d(S/I_i)).$$

Therefore, we conclude that theorem 3.10 and 3.11 hold for M, provided that we can prove that they hold for all modules of the form S/I, where I is a T–stable prime ideal in $S = S(E^*)$, and $d(S/I) \le d(M)$.

In other words, by the preceding argument we have reduced the proof to the particular case $M = S/I$, with I a T–stable prime ideal. To prove the theorem for this case, pick some weight vector $0 \ne f \in M_{-\mu}$ of strictly negative weight $-\mu$; this is possible: in fact, we can take any weight vector in S/I except the scalars. Then $q_M = \mu^{-1} q_{M/fM}$ by the lemma (3.12), and $d(M/fM) < d(M)$, so q_M will be positive, since $q_{M/fM}$ is by induction hypothesis. Moreover, since

$$\deg q_{M/fM} = r{-}d(M/fM)$$

by induction, the lemma gives also

$$\deg q_M = (\deg q_{M/fM}) - 1 = r{-}d(M/fM){-}1 = r{-}d(M) .$$

This completes our inductive proof of theorem 3.10 and 3.11. Q.e.d.

3.14 Determination of character polynomials by supports.

If $\mathcal{V} \subset E$ is a t–stable closed subvariety, then the structure sheaf $F = \mathcal{O}(\mathcal{V})$
is T–equivariant, and so has a character polynomial, which we also denote
$q_{\mathcal{V}} = q_{\mathcal{O}(\mathcal{V})}$, for short. If F is an arbitrary T–equivariant coherent sheaf on E,
then its support in E is clearly a T–stable closed subvariety. Recall our notation
for the *supporting cycle* of F: This is the formal linear combination

(*) $\mathrm{supp}(F) = \sum_{\mathcal{V}} m_{\mathcal{V}}(F)[\mathcal{V}]$,

extended over the irreducible components \mathcal{V} of the support (notation 1.4).

Proposition: *Let F be a T–equivariant coherent sheaf on E, let d denote the
codimension of its support in E, and let (*) be its supporting cycle. Then*

$$q_F = \sum_{\mathrm{codim}\ \mathcal{V}=d} m_{\mathcal{V}}(F)q_{\mathcal{V}} .$$

In particular, the character polynomial of F is completely determined by its
supporting cycle.

Proof: We let $M = \Gamma(E,F)$ and chose prime ideals $I_1,...,I_\ell$ as in lemma 3.4. Let
\mathcal{V}_j denote the support of S/I_j in E; then $q_{S/I_j} = q_{\mathcal{V}_j}$. We already knew that

$$q_F = \sum_j{}' q_{S/I_j} = \sum_j{}' q_{\mathcal{V}_j} ,$$

where \sum' is summation over those j for which q_{S/I_j} has minimal degree (cf.
3.13). But from theorem 3.10, we now know in addition that

$$\deg P_F = d \ , \ \deg q_{\mathcal{V}_j} = d_j$$

for $j = 1,...,\ell$, where d_j denotes the codimension of \mathcal{V}_j in E. We conclude that

(**) $q_F = \sum_{d_j=d} q_{\mathcal{V}_j} .$

On the other hand, it is clear that the V_j of codimension $d_j = d$ are the irreducible components of the support of F of minimal codimension. Now (**) gives the proposition. $\hspace{4cm}$ Q.e.d.

Corollary: *Let* $0 \to M_1 \to M \to M_2 \to 0$ *be a short exact sequence of* T*–equivariant finitely generated* $S(E^*)$*–modules. If* $d(M_1) = d(M_2)$, *then*

$$q_M = q_{M_1} + q_{M_2} .$$

If $d(M_i) < d(M_j)$ $(\{i,j\} = \{1,2\})$, *then*

$$q_M = q_{M_j} .$$

3.15 The theory of Hilbert–Samuel polynomials as a special case [AC], [AM].

Let us consider the case $T = G_m = k^*$ more closely. The character group $X^*(T)$ is infinite cyclic, with generator the identity character t (given by $a \longmapsto a$) of $T = G_m$. We identify $R(T)$ with the ring of Laurent polynomials in one variable t. The order on $X^*(T) = \mathbb{Z}$ is the usual one. We have: $R(T) = k[t,t^{-1}]$, the augmentation ideal I_T is the principal idea generated by $t-1$, and the I_T–adic degree of a Laurent polynomial $F(t)$ is the order of vanishing of $F(t)$ at $t = 1$. To summarize:

Lemma: *The completed representation ring of the multiplicative group is the ring of formal power series in the variable* $t-1$,

$$\hat{R}(G_m) = k[[t-1]] ,$$

the γ*–degree of a power series being its order of vanishing at* $t = 1$.

A T–action on a module M is now just a \mathbb{Z}–grading $M = \underset{i \in \mathbb{Z}}{\oplus} M_i$ (M_i is the subspace of weight i, on which a acts by a^i), and M will admit a formal chardimensional, the formal character being known as the *Poincaré series*

$$ch(M) = \underset{i}{\Sigma} (\dim M_i) \cdot t^i$$

in this case. Now let $d_1,...,d_r \in \mathbb{Z}$ denote the weights of our linear action of $T = G_m$ on the r–dimensional vector space E. By our assumption, they are all positive. Recalling 3.3, we find that now

$$\Delta(E) = (1-t^{d_1})...(1-t^{d_r}) ,$$

so

$$ch(S(E^*)) = \Delta(E^*)^{-1} = (1-t^{-d_1})^{-1}...(1-t^{-d_r})^{-1} .$$

Now let **F** be a T–equivariant coherent \mathcal{O}_E–module, and $M = \Gamma(E,F)$ the corresponding graded $S(E^*)$–module. Then our result in proposition 3.7 and theorem 3.10, specialized to this case, yields in particular the following facts about Poincaré series.

Corollary: *The Poincaré series of the \mathbb{Z}–graded module* M *(as above) takes the form of a rational function*

$$ch(M) = \frac{P(t)}{(1-t^{-d_1}) ... (1-t^{-d_r})} ,$$

where $P(t)$ *is a Laurent polynomial, whose order of vanishing at* $t = 1$ *equals* $r-d(M)$, *the codimension of the support of* M *in* E.

For example, let T be the group of homotheties on E. Then $d_1 = d_2 = ... = d_r = 1$, so the Poincaré series equals

$$ch(M) = \frac{P(t)}{(1-t^{-1})^r} = (1-t^{-1})^{-d(M)} Q(t) ,$$

where $Q(t)$ is a polynomial with $Q(1) \neq 0$.

Remark: These are essentially classical results of Hilbert and Samuel in commutative algebra, for expositions of which we refer to e.g.: Bourbaki, [AC] VIII, §4, n°3, Théorème 2, and n°4, Théorème 3, and also [AM], theorems 11.1 and 11.14, or [Sr], under the headlines of "degree of the Hilbert–Samuel functions", resp. "Hilbert functions" and "dimension theory". Note, however, that ch(M) has

$(1-t^{-1})^r$ in its denominator, instead of the usual $(1-t)^r$.

3.16 Restriction to one–parameter subgroups.

Let $\alpha_1,...,\alpha_\ell$ be generators of the monoid Ω of ≥ 0 weights. Our positivity assumption on the T module E means that its weights belong to $\Omega\backslash\{0\}$, and furthermore for any T–equivariant $S(E^*)$–module M, there exists some $\lambda \in \Lambda$ such that

$$ch(M) = \sum_{-\mu \in \lambda+\Omega} (\dim M_\mu)e^\mu \qquad (*)$$

for each T–equivariant $S(E^*)$–module. We introduce the exponentials $t_1 := e^{\alpha_1},...,t_\ell := e^{\alpha_\ell}$. So we may consider the formal character as a formal Laurent power series in $t_1,...,t_\ell$ as mentioned in 3.2, that is

$$ch(M) = F(t_1,...,t_\ell) = \sum_{m_1,...,m_\ell \geq -k} a_{m_1,...,m_\ell} t_1^{m_1} t_2^{m_2} ... t_\ell^{m_\ell}$$

with $m_1,...,m_\ell$ integers bounded below by some $-k$.

Now we consider a *one parameter subgroup* $\psi : G_m \rightarrow T$. We may describe ψ in terms of the *perfect pairing* $X_*(T) \times X^*(T) \rightarrow \mathbb{Z}$ of the (rank ℓ free abelian) groups $X_*(T)$, resp. $X^*(T)$ of all one parameter subgroups, resp. characters of T (see e.g. [S2], 2.5.12) as follows: For a character $\chi \in X^*(T)$, we define $<\chi,\psi>$ as the integer n such that the restricted character $\psi^*(\chi)$ of G_m is a $\longmapsto a^n$ ($a \in G_m$). We denote $\chi_1,...,\chi_\ell$ the characters of T corresponding to $\alpha_1,...,\alpha_\ell$. Then ψ is uniquely determined by the ℓ integers

$$n_i := <\chi_i,\psi> \qquad (i = 1,...\ell) .$$

Definition: ψ is called *positive*, if all $n_1,...,n_\ell$ are *positive*.

Lemma: *The* G_m*-module obtained from* M *by restriction to a positive one parameter subgroup* ψ *admits a formal character* $\psi^* \mathit{ch}(M)$. *If*

$\mathit{ch}(M) = F(t_1,...,t_\ell)$ *as above, then*

(*) $$\psi^* \mathit{ch}(M) = F(t^{n_1},...,t^{n_\ell}).$$

Here $t = e^{-\alpha}$ denotes the character $a \longmapsto a$ of G_m .

The lemma is obvious for a one–dimensional module $M = k_\chi$ of character

$\mathit{ch}(M) = [\chi] = t_1^{m_1} ... t_\ell^{m_\ell}$, since then

$$\psi^* \mathit{ch}(M) = t^{<\chi,\psi>} = t^{n_1 m_1 + ... + n_\ell m_\ell} = (t_1^{n_1})^{m_1} ... (t_\ell^{n_\ell})^{m_\ell},$$

and it extends linearly to arbitrary modules M of finite dimension, i.e. to

polynomials F. For the extension to arbitrary bounded formal characters $\mathit{ch}(M)$,

i.e. to Laurent *power series* F, we need the positivity assumption on ψ. This will

guarantee the existence and boundedness of the formal character of M, considered

as a G_m–module by restriction, and then formula (*) follows as in the

finite–dimensional case.

Proposition: Let ψ be a positive one parameter subgroup of T. Let

$\psi^* : K_T(E) \longrightarrow K_{G_M}(E)$ denote restriction. Let $[M] \in K_T(E)$ denote the class

determined by some T–equivariant $S(E^*)$ module M, and d its γ degree. Then

 a) γ–deg $\psi^*[M] \geq d = \gamma$–deg $[M]$.

 b) γ–deg $\psi^*[M] = d$ if and only if

$$q_M^T (y) \neq 0, \text{ at } y := d\psi(1) \in t.$$

 c) *If such is the case, then the character polynomial of the "restricted"*

module is obtained by restriction of the character polynomial, i.e., $q_M^{G_m} = \psi^* q_M^T$.

If we write $q_M^T = f(\omega_1,...,\omega_\ell)$ *as a polynomial in the fundamental weights, then*

$$\psi^* q_M^T = f(n_1,...,n_\ell) \, \alpha^d$$

with $n_i = <\chi_i, \psi>$ *as above.*

Proof: Referring to 3.7 and the lemma above, we may describe the restriction map ψ^* explicitly by means of the following commutative diagram,

$$
\begin{array}{ccccc}
K_T(E) & \xrightarrow{\iota_T^*} & R(T) & \xrightarrow{\sim} & k[t_1, t_1^{-1}, \ldots, t_\ell, t_\ell^{-1}] \\
\psi^* \downarrow & & \psi^* \downarrow & & \psi^* \downarrow \\
K_{G_m}(E) & \xrightarrow[\iota_{G_m}]{\sim} & R(G_m) & \xrightarrow{\sim} & k[t \cdot t^{-1}]
\end{array}
$$

where the vertical arrows will all be denoted ψ^*, but the last one is explicitly described as a specialization map of Laurent polynomials as follows:

$$\psi^* P(t_1, \ldots, t_\ell) = P(t^{n_1}, \ldots, t^{n_\ell}) .$$

Hence if $[M]$ is represented by the multi–variable Laurent polynomial $P(t_1, \ldots, t_\ell)$, then $\psi^*[M]$ is represented by the one–variable Laurent polynomial $P(t^{n_1}, \ldots, t^{n_\ell}) =: P'(t)$. Now recalling $t_i = e^{\alpha_i}$, and $t = e^{\alpha}$, we may develop these polynomials into power series in the fundamental weights $\alpha_1, \ldots, \alpha_\ell$ resp. α. If we write

$$P(t_1, \ldots, t_\ell) = \sum_{j \geq d} q_j(\alpha_1, \ldots, \alpha_\ell) ,$$

with q_j a homogeneous polynomial of degree j in $\alpha_1, \ldots, \alpha_\ell$ and $q_d \neq 0$, then the character polynomial of M with respect to T is

(1)
$$q_M^T = q_d(\alpha_1, \ldots, \alpha_\ell) ,$$

whereas that of M with respect to ψ (i.e., $q_M^{G_m}$) is the lowest term of the power series. Now

(2)
$$P'(t) = P(t^{n_1}, \ldots, t^{n_\ell}) = P(e^{n_1 \alpha}, \ldots, e^{n_\ell \alpha}) = \sum_{j \geq d} q_j(n_1 \alpha, \ldots, n_\ell \alpha) =$$

$$\sum_{j \geq d} q_j(n_1, \ldots, n_\ell) \alpha^j,$$

and the statements of the proposition are then obvious. In particular

$$q_{\mathbf{M}}^{\mathbf{G}_m} = q_d(n_1,...,n_\ell) \; \alpha^d,$$

if this is non–zero. Q.e.d.

Corollary: *For a given* T–*equivariant module* M, *restriction preserves* γ–*degree, that is equality holds in* a) *above, for "almost all" positive one parameter subgroups* ψ, *in the sense that the exceptional set is contained in a Zariski closed subset of* $X_*(T) \otimes_{\mathbb{Z}} k.$

Recall that $X_*(T)$ denotes the group (isomorphic to \mathbb{Z}^ℓ) of all one parameter subgroups ψ of T.

Remarks: 1) Equality holds in a) above even *all* positive one parameter subgroups ψ by theorem 3.10. In fact, theorem 3.10 says that the γ–degree of M equals the codimension of its support, and hence is completely independent of the torus action under consideration, provided, that this action has only positive weights on E, which is guaranteed in the present situation for our one parameter subgroup ψ since it is assumed *positive*.

2) The reason for stating the proposition in the weaker form above which we have easily proved directly (independent of 3.10), is that it provides an *alternative* method of *proof* for theorem 3.10: In fact, the corollary above reduces the proof of the theorem to the special case $T = G_m$ of a one–dimensional torus. It seems to us that even this case is not fully covered by the existing literature (cf. e.g. [Bo], [AM], [Sr], [Sm], theorem 5.5). Let us give therefore another full proof in 3.18, which is essentially based on a lemma about real power series (3.17), and which might have some independent interest.

3.17 A lemma on the growth of coefficients of a power series.

Let $(H_j)_{j \in \mathbb{Z}}$ be a sequence of *non–negative* real numbers, with a minimal positive term H_{j_0} (that is $H_j = 0$ for $j < j_0$). Let us define its *order of growth* as the infimum of all real numbers $\gamma \geq 0$ such that $\sum_{j \leq n} H_j = nO(n^{\gamma})$ as n tends to infinity. (For example, the order of growth of a polynomial function is its degree.) Now we assume that the Laurent power series

$$H(t) = \sum_j H_j t^j \qquad (1)$$

is a *rational function of the form*

$$H(t) = P(t)/ \prod_{i=1}^{r} (1-t^{m_i}) , \qquad (2)$$

where $P(t)$ is a Laurent *polynomial*, and $r, m_1, ..., m_r$ are positive integers.

Lemma: *The order of growth of* (H_j) *is* $r-d-1$ *where* d *is the order of vanishing of* $P(t)$ *at* $t = 1$.

Proof: We first note that with $H(t)$ also $(1+t+t^2+...+t^{s-1})H(t)$ has non–negative coefficients and satisfies all assumptions of the lemma (for all $s \geq 0$); moreover, since all its coefficients are the sums of s successive coefficients $H_{j-s+1} + ... + H_j$ of $H(t)$, it is obvious that the order of growth of the coefficients is not changed by multiplication with such a polynomial $(1+t+t^2+...+t^{s-1})$. By applying this observation r times, we conclude that the power series

$$U(t) = H(t) \prod_{i=1}^{r} (1+t+t^2 + ... \; t^{m_i-1})$$

has also non–negative coefficients of the same growth as $H(t)$. On the other hand, equation (2) gives now

$$U(t) = P(t) / (1-t)^r .$$

Consequently, we have reduced the proof of the lemma to the special case where all

degrees are one, $m_1 = m_2 = \ldots = m_r = 1$. In this case, the proof of the lemma is easy by use of the binomial series

$$(1-t)^{-r} = \sum_{i=0}^{\infty} \binom{-r}{i}(-t)^i = \sum_{i=0}^{\infty} \binom{r+i-1}{r-1} t^i \ ,$$

see [Bo]. Q.e.d.

3.18 An alternative proof of theorem 3.10.

As mentioned before, in view of Corollary 3.16, it suffices to prove the theorem for the special case $T = G_m$, acting on E with positive weights m_1,\ldots,m_r. Let F and $M = \Gamma(E,F)$ as in 3.10, and consider the formal character or Poincaré series of M, denoted

$$ch(M) = \sum_j H_j t^j = H(t) \ ,$$

($H_j = \dim M_{-j}$). We know from 3.7, 3.15 (independently of theorem 3.10) that this series is a rational function of the form considered above, in 3.17 (2). Hence by lemma 3.17, we know the order of growth of (H_j) to be $r-d-1$, where $r = \dim E$, and $d = \gamma-\deg[M]$ (use lemma 3.15). Now the proof is completed by the following

Lemma: *The order of growth of* (H_j) *is equal to* *dim supp* $F-1$.

This lemma is easily proven via an analogue, for orders of growth, of lemma 3.11.

 Q.e.d.

§4. Equivariant characteristic classes of orbital cone bundles.

In this chapter, G is a connected semisimple algebraic group over k, and T a maximal torus in G. We use the notations introduced in 1.5. So, in particular, U is a maximal unipotent subgroup normalized by T, and B = TU = UT is a fixed Borel subgroup; $\mathfrak{g},\mathfrak{b},\mathfrak{t},\mathfrak{u}$ are the Lie algebras of G,B,T,U, etc.

4.1. Borel picture of the cohomology of a flag variety [Bo], [H1].

The purpose of the theorem below is to recall, and simultaneously to rephrase in an equivariant K–theory language convenient for our present paper, the Borel picture of the cohomology ring $H^*(X)$ of the flag variety $X = G/B$. This picture describes $H^*(X)$ as a quotient of the representation ring $R(T)$ by a certain ideal defined as follows. Note that the Weyl group W of G relative to T acts on $R(T)$, stabilizing the augmentation ideal I_T (notation 2.11); so let $I_T^W = I_T \cap R(T)^W$ denote the subspace of W invariants. Then the ideal mentioned above is the one generated by I_T^W.

Theorem (Borel picture): *We have a commutative diagram of canonical ring homomorphisms*

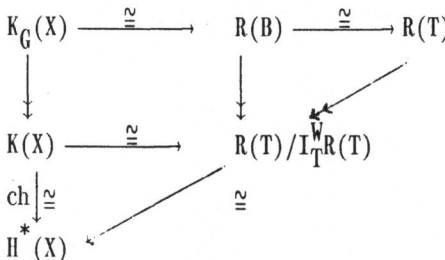

Here the map from $K_G(X)$ onto $K(X)$ forgets the G–action (2.1), and the isomorphism of $K_G(X)$ onto $R(B)$ is given by restriction to the base point (2.4(1)). The link from $K(X)$ to $H^*(X)$ is of course made by the Chern character

ch (an isomorphism, cf. 1.5), and that from $R(B)$ to $R(T)$ is given by restriction from B to T (2.1). Observing that U acts trivially on each finite dimensional irreducible (hence one–dimensional) B–module, we conclude that this restriction homomorphism $R(B) \to R(T)$ is also an isomorphism. For further information and more classical formulations of the Borel picture, we refer to e.g. Borel [Bo], or Hiller [H1], III, theorem 4.1.

4.2 Description in terms of harmonic polynomials on a Cartan subalgebra.

The isomorphism ψ of the completed representation ring $\hat{R}(T)$ with the formal power series ring $\hat{S}(t^*)$ (cf 3.1) is canonical, hence W–equivariant; it maps the ideal generated by I_T^W onto the ideal generated by

$$J_T^W := S(t^*)^W \cap t^* S(t^*) \,,$$

that is the W–invariant polynomial functions on t vanishing at 0. Therefore ψ induces a diagram of isomorphisms

$$
\begin{array}{ccccc}
\hat{R}(T) & \longrightarrow & \hat{R}(T)/\hat{R}(T)I_T^W & \xleftarrow{\ \simeq\ } & R(T)/R(T)I_T^W \\
\downarrow{\simeq} & & \downarrow{\simeq} & & \downarrow{\simeq} \\
\hat{S}(t^*) & \longrightarrow & \hat{S}(t^*)/\hat{S}(t^*)J_T^W & \longleftarrow & S(t^*)/S(t^*)J_T^W
\end{array}
$$

Furthermore, we recall that the residue class map of $S(t^*)/S(t^*)J_T^W$ admits a natural section, given by the graded subspace $S(t^*)^\natural \subset S(t^*)$ of all W–*harmonic polynomials*. In conclusion, Borel's description of $H^*(X)$ as formulated in 4.1 also says that

$$H^*(X) \simeq R(T)/R(T)I_T^W \simeq S(t^*)/S(t^*)J_T^W \simeq S(t^*)^\natural$$

canonically. These isomorphisms clearly respect gradings and W actions. Thus we reobtain the following formulation of the Borel picture, which is perhaps more familiar to some readers:

Corollary (loc. cit.): *For all degrees* $d \geq 0$, *we have canonically*

$$H^{2d}(X) \cong S^d({t^*})^\natural.$$

Here the right—hand side denotes the space of harmonic polynomials which are homogeneous of degree d. If we add that the cohomology of X vanishes in all odd degrees, then the above statement completes the *description of* $H^*(X)$ *as a graded ring with* W *action.*

Remark: Let us recall our convention, already made in 1.6, that whenever convenient, we identify a cohomology class on X with a harmonic polynomial on t, by means of the above isomorphism.

4.3 Equivariant K—theory on T^*X.

We shall now return to the study of certain conical subvarieties in — resp. sheaves on — the cotangent bundle T^*X of the flag variety X, which has been our main object in §1 already. We now observe that T^*X is G—equivariant as a vector bundle over X, and that the subvarieties and sheaves in question are also G—equivariant. This will allow us to study them by means of calculations in the ring $K_G(T^*X)$. This study will amount in a certain sense to "lifting" the geometrical investigations of our §1 to the more refined level of *equivariant* K—theory in the present chapter, and will finally lead to improvements of the results in §1.

The purpose of the following proposition is to provide a fairly explicit description of the ring $K_G(T^*X)$, as well as a method for performing actual calculations. We recall that the cotangent space of X at the base point x = {B} is identified with the "co—isotropy space" at x, that is to say $(T^*X)_x = b^\perp = u$, and then T^*X is identified with the associated fibre bundle $T^*X = G \times^B u$ (notations 1.5). Consider the following diagram of inclusion maps:

Here σ resp. j denote the inclusion of X resp. u into T^*X as the zero section resp. the fibre at x.

Proposition: *We have a commutative diagram of canonical isomorphisms of λ rings:*

$$
\begin{array}{ccccc}
K_G(T^*X) & \xrightarrow{\ j_B^*\ } & K_B(u) & \longrightarrow & K_T(u) \\
\Big\downarrow{\sigma_G^*} & & \Big\downarrow{\iota_B^*} & & \Big\downarrow{\iota_T^*} \\
K_G(X) & \xrightarrow{\ i_B^*\ } & R(B) & \xrightarrow{\ \rho\ } & R(T)
\end{array}
$$

Here all maps are given by restriction to a smaller group and/or to a subvariety.

Proof: All these restriction homomorphisms are clearly λ–ring homomorphisms, so it suffices to check bijectivity. For the two horizontal arrows j_B^*, i_B^*, this is a consequence of the G–homogeneity of X (cf. 2.4(1)). Also, the horizontal arrow ρ is an isomorphism, observed already in 4.1, and the vertical arrow ι_T^* is an isomorphism by proposition 3.7. The order on $X^*(T)$ is the one for which the monoid of ≥ 0 elements is generated by the simple roots $\alpha_1,...,\alpha_\ell$. Identifying R(B) with R(T), we may repeat the proof of proposition 3.7 almost word by word to construct an inverse map of ι_B^* from $R(T) = R(B)$ to $K_B(u)$. Now the bijectivity of the remaining two arrows follows by commutativity of the diagram.

Q.e.d.

Remark: Alternatively, one may invoke Thomason's general result [T1], Theorem 4.1 (cf. 2.9(2)) to conclude directly that all three vertical arrows ι_T^*, ι_B^*, σ_G^* are

isomorphisms. However, for the convenience of the reader, we are avoiding here the use of this difficult result.

4.4 Restriction to a fibre of T^*X

Let us now explain in more detail the use of the proposition above for computational purposes. Let us compute for instance the class in $K_G(T^*X)$ determined by some G–equivariant coherent sheaf F on T^*X. To identify this class $[F]$ as an element in $R(T)$ by means of the proposition, we proceed as follows: We restrict F to a sheaf j^*F on the fibre u (which is affine!), then take its module of global sections $M = \Gamma(u, j^*F)$ and the formal character $ch(M)$ in $\hat{R}(T)$ (3.2). Then the desired element in $R(T)$ is given by the explicit formula for the map ι_T^* given in 3.7, that is to say:

Corollary 1: $\qquad\qquad \iota_T^* \, j_B^*[F] = \Delta(u^*) \, ch \, \Gamma(u, j^*F).$

Here $\Delta(u^*)$ is the product $\Pi(1-e^{-\alpha})$, expanded over all positive roots α (relative \mathfrak{b} and \mathfrak{t}).

Corollary 2: Let F *denote a* G–*equivariant coherent sheaf on* T^*X. *Then the* γ–*degree of its class* $[F]$ *in* $K_G(T^*X)$ *equals the codimension of the support of* F *in* T^*X.

Proof: This follows from theorem 3.10 by means of proposition 4.3.

4.5 Definition of equivariant characteristic classes.

Let F be a G–*equivariant* coherent sheaf on T^*X, whose support has codimension d in T^*X. Then its *equivariant characteristic class* $Q_G(F)$ is a

homogeneous polynomial of degree d on \mathfrak{t}, which is defined as follows. First of all, F determines a class $\sigma_G^*[F]$ in $K_G(T^*X)$. By "homological intersection with the zero section", it determines a class $\sigma_G^*[F]$ in $K_G(X)$, which we may consider as a power series on \mathfrak{t}, by means of the identifications

$$K_G(X) \cong R(T) \subset \hat{R}(T) \cong \hat{S}(\mathfrak{t}^*)$$

explained in 4.3 resp. 3.1. Then we *define* $Q_G(F)$ as the lowest degree term of this power series:

$$Q_G(F) := \text{gr } i_B^* \sigma_G^*[F] .$$

Proposition: *The polynomial $Q_G(F)$ may be computed from a formal character by means of the formula*

$$Q_G(F) = \text{gr } \Delta(u^*)\text{ch } \Gamma(u, j^* F)$$

In particular, it is homogeneous of degree d.

Proof: The first statement follows from corollary 4.4.1, in view of the commutativity of the diagram in proposition 4.3. The second statement follows from corollary 4.4.2, so is a consequence of theorem 3.10. Q.e.d.

Remark. Using the terminology and notations of character polynomials as defined in 3.9, we may also state the proposition this way: As a polynomial on \mathfrak{t}, the equivariant characteristic class of F coincides with the character polynomial of its restriction to a fibre of T^*X, that is to say

$$Q_G(F) = q_{j^* F}^T .$$

4.6. Comparison to the characteristic classes defined in §1.

Retaining the notations of 4.5, let us recall that we already defined a *characteristic class* $Q(F)$ in chapter 1, as an element in $H^{2d}(X) \cong S^d(\mathfrak{t}^*)^\natural$ (see 1.2

and 4.2). Let us denote $p \longmapsto p^{\natural}$ the projection of a polynomial p on t onto its uniquely determined harmonic part p.

Proposition: *The characteristic class* $Q(F)$ *may be computed from the equivariant one* $Q_G(F)$ *by taking the harmonic part,*

$$Q(F) = Q_G(F)^{\natural}$$

Proof: From the "Borel picture" (theorem 4.1, 4.2) we obtain the following commutative diagram by passing to the associated graded level (terminology and notations as in 1.2, 1.4):

$$
\begin{array}{ccccc}
\mathrm{gr}_d K_G(X) & \xrightarrow{\ \sim\ } & \mathrm{gr}_d R(T) & \xrightarrow{\ \sim\ } & S^d(t^*) \\
\downarrow & & \downarrow & & \downarrow{\scriptstyle \natural} \\
\mathrm{gr}_d K(X) & \xrightarrow{\ \sim\ } & \mathrm{gr}_d R(T)/R(T)I_T^W & \xrightarrow{\ \sim\ } & S^d(t^*)^{\natural}
\end{array}
$$

Here the left vertical arrow is given by *forgetting* the G–action, while the right one is taking harmonic parts. So the diagram says that forgetting the G action in graded K groups corresponds to taking harmonic parts of the corresponding polynomials on t. Now note that $Q(F)$ was *defined* as an element in

$$\mathrm{gr}_d K(X) \cong H^{2d}(X) \cong S^d(t^*)^{\natural},$$

and $Q_G(F)$ is an element in

$$\mathrm{gr}\, K_G(X) \cong \mathrm{gr}\, R(T) \cong S(t^*)$$

by definition. The only delicate point is that $Q_G(F)$ and $Q(F)$ should occur in the same degree, that is in degree d. But this was ensured by theorem 3.10 (cf. 4.4.2). Now it is clear that $Q(F)$ is obtained from $Q_G(F)$ by forgetting the G–action, and so the proposition follows from the above commutative diagram.

Q.e.d.

Complementary remarks.

1) If $\mathcal{K} \subset T^*X$ is a G stable closed subvariety, then we define its equivariant characteristic class by

$$Q_G(\mathcal{K}) := Q_G(\mathcal{O}_\mathcal{K}) \ .$$

Since we may define $Q(\mathcal{K})$ analogously as $Q(\mathcal{O}_\mathcal{K})$ (see 1.4), we see that also

$$Q(\mathcal{K}) = Q_G(\mathcal{K})^\natural.$$

2) Let us recall that it was a delicate question in §1, whether or not the characteristic class $Q(\mathcal{V})$ of a given subvariety $\mathcal{V} \subset T^*X$ vanishes. So let us point out here for the case of a G–stable subvariety $\mathcal{V} \neq \emptyset$ that $Q_G(\mathcal{V})$ is always nonzero by definition, and that the delicate question is, whether or not the corresponding polynomial has *nonzero harmonic part*. In our case of main interest, that is for \mathcal{V} an orbital cone bundle, it actually turns out that even the equivariant class itself is already harmonic, so that $Q(\mathcal{V}) = Q_G(\mathcal{V})$ in this case (so in particular $Q(\mathcal{V}) \neq 0$).

3) Let us point out that the *equivariant characteristic class* $Q_G(F)$ may be described as a class in the *equivariant cohomology group* $H_G^{2d}(X)$. We use the standard definition of equivariant cohomology, that is $H_G^*(X) = H^*(X \times^G EG)$ (see [Bo2], chapter 4). It is also easy to define a purely algebraic equivariant de Rham cohomology, by viewing $X \times^G EG$ as a simplicial algebraic variety over k, and taking the hypercohomology of the de Rham complex (which is a simplicial complex of sheaves) (see [D1] or [Fr] for general notions about simplicial varieties and simplicial sheaves). In our case, where $X = G/B$, we have $H_G^{2d}(X) \xrightarrow{\sim} S^d(t^*)$ (and $H_G^k(X) = 0$ for k odd). Also, the restriction to the zero section induces an isomorphism $H_G^*(T^*X) \xrightarrow{\sim} H_G^*(X)$. There is a general equivariant Chern character $ch_G = ch : K_G(-) \to H_G^{even}(-)$, where $H_G^{even}(-)$ is the *product* of all $H_G^{2i}(-)$. For the flag variety X, we have a commutative diagram

where the lower line is defined in 3.1. It follows that ch maps $K_G(X)_d$ to $\prod\limits_{i \geq d} H_G^{2i}(X)$. So we may give an alternative geometric definition of $Q_G(F)$ as follows. Take F as in 4.5, then $Q_G(F)$ is the component of ch($[F]$) in $H_G^{2d}(X)$. Remembering the isomorphism $H_G^{2d}(X) \stackrel{\sim}{\to} S^d(t^*)$ we get the same class as in 4.5. The commutative diagram used in the above proof of Proposition 4.6 may be completed as follows

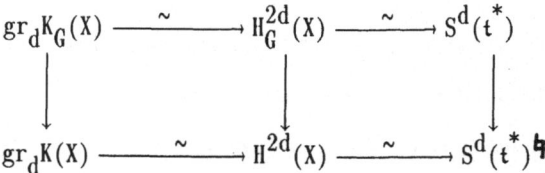

We will not try to give a purely homological definition of $Q_G(V)$, in the style of 1.3, because of the lack of references concerning equivariant Borel–Moore homology.

4) The cotangent bundle T^*X may be viewed as a G–equivariant vector bundle on X. Since the equivariant Chern classes $c_i(T^*X) \in H_G^{2i}(X)$ are defined, we may, by the usual procedure [Hi], §10.1, introduce the *Todd class* $Td(T^*X)$. To compute it, we freely identify $H_G^{even}(X)$ with $\hat{S}(t^*)$, as in the previous remark.

We have $\sum\limits_{i=1}^{n} c_i(T_X^*)x^i = \prod\limits_{\alpha>0} (1+\alpha x)$, hence $Td(T_X^*) = \prod\limits_{\alpha>0} (\frac{\alpha}{1-e^{-\alpha}}) = \frac{c_n(T_X^*)}{\Delta(u^*)}$
(this is an equality in the fraction field of $\hat{S}(t^*)$).

Now let F be a G–equivariant coherent sheaf on T^*X, let $M = \Gamma(u,j^*F)$. Then the formula for ch(M) in 4.4 may be rewritten as:

$$ch(M) = \frac{Td(T_X^*) \cdot Q_G(F)}{c_n(T_X^*)}$$

4.7. Equivariant characteristic classes of orbital cone bundles.

We are now ready to state some of the key results of the present chapter.

Theorem 1. *Let \mathcal{K} be an orbital cone bundle in T^*X (terminology 1.7). Then its equivariant characteristic class $Q_G(\mathcal{K})$ is a harmonic polynomial.*

In other words, we have $Q_G(\mathcal{K}) = Q(\mathcal{K})$ in this case.

Theorem 2: *Let $\mathcal{K}_1,...,\mathcal{K}_r$ be the orbital cone bundles for some nilpotent orbit \mathcal{O}_u (notations as in theorem 1.8). Then the classes $Q_G(\mathcal{K}_1),...,Q_G(\mathcal{K}_r)$ span a Weyl group submodule.*

Theorem 3: *The corresponding representation of* W *is isomorphic to Springer's irreducible representation ρ_u (notation as in theorem 1.8).*

We note that theorems 2 and 3 above would be immediate consequences of theorem 1.8, if we would assume theorem 1. However, our proof proceeds in a different logical order. Logically, the proof of theorem 2 comes first; it will be completed in 4.13. But let us show here first how to derive theorems 1 and 3 from theorem 2.

Theorem 2 implies theorem 3: Let M resp. N denote the linear span of $Q_G(\mathcal{K}_1),...,Q_G(\mathcal{K}_r)$ resp. of $Q(\mathcal{K}_1),...,Q(\mathcal{K}_r)$. By proposition 4.6, M is mapped onto N by the projection onto harmonic parts, which is a W–module homomorphism. Its restriction to M must be a linear isomorphism by 1.8a). So assuming that M is a W–submodule, we conclude that it carries the same W–representation as N, which is ρ_u by theorem 1.8c).

Q.e.d.

Theorems 2 and 3 imply theorem 1: Suppose not all $Q_G(\mathcal{K}_i)$ were harmonic. With notations as before, this means M ≠ N, and so M,N are two different copies

of the irreducible representation ρ_u in $S^{d_u}(t)$ (notation 1.8). However, it is known that Springer's representation ρ_u occurs only *once* in (the lowest possible) degree d_u, see [BM1], Corollaire 4. Hence we must have $M = N$.

<div align="right">Q.e.d.</div>

The *proof of theorem* 2 will consist in verifying directly the following much more precise statement:

Theorem 2ʹ : *Even the equivariant characteristic classes* $Q_G(\mathcal{K}_1),...,Q_G(\mathcal{K}_r)$ *satisfy the Hotta transformation formulas* (cf. 1.15).

This is our version of a result essentially due to Joseph [J1], 3.1, and reinterpreted by Hotta [Ho], 3.4. In addition to proving theorem 2, this stronger theorem 2ʹ will simultaneously prove theorem 1.15 (see 4.14 below), and hence reprove also the main results of Hotta's work [Ho], [Ho2]. The proof will be given in sections 4.8 to 4.13. The crucial final part of the proof (cf. 4.13) is essentially Joseph's so that we could refer partly to [J1]. However, for convenience of the reader, we prefer to give here an essentially self–contained, complete proof.

4.8 Comparison with Joseph's notion of "characteristic polynomial"

The purpose of this section is to link up terminology and notations of Joseph [J1] with our present language. The main point to be made is that our notion of "equivariant characteristic classes" essentially coincides with Joseph's notion of "characteristic polynomial", providing a geometric reinterpretation for it. Let us give a few more explanations, which should help the reader to verify the coincidence.

We shall identify here polynomials on t^* with polynomials on t by means of the Killing form, that is by the isomorphism $t \cong t^*$, $t \longmapsto t^*$, characterized by

$\mu(t) = (\mu, t^*)$. Let $C \subset u$ be a B–stable closed cone, and let $\mathcal{K} = G \times^B C \subset T^* X$
denote the corresponding G–stable closed cone bundle.

Proposition: *Considered as a polynomial on* $t = t^*$, *our equivariant characteristic*
class $Q_G(\mathcal{K})$ $(= q_C^T$ *by 4.5) coincides with Joseph's "characteristic polynomial"*
p_C, *as defined in* [J1], *2.4.*

1) Let us first recall Joseph's definition (loc.cit.). Let M denote the
T–equivariant $S(u^*)$–module $M = \mathcal{O}(C)$. Let $\nu \in t^*$ denote an integral weight,
which is assumed dominant and regular, that is $(\alpha, \nu) > 0$ for all positive roots α;
ν will eventually be considered as a variable. Now Joseph considers the "Poincaré
series"

$$R_M(t, \nu) := \sum_{\mu \in \Lambda} (\dim M_\mu) t^{(\mu, \nu)}$$

(notation as in [J1]), which is a formal power series in one variable t, and which
depends on ν as a parameter. Next he studies the leading coefficient of the
Hilbert–Samuel polynomial of $R_M(t, \nu)$, considered as a function of ν, denoted
$r_M(\nu)/d(M)!$, and he finds that the function

$$p_M(\nu) := r_M(\nu) \cdot \prod_{\alpha > 0} \alpha(\nu) \tag{1}$$

is given by a polynomial p_M on t^*, homogeneous of degree
$d := \mathrm{codim}_u C = n - d(M)$. This polynomial is denoted $p_C := p_M$, and is called
the "characteristic polynomial of C" in loc.cit.

2) Next let us rewrite Joseph's procedure in our present language. We
consider the restrictions of the T–action on our T–equivariant module
$M = \mathcal{O}(C) = \Gamma(C, \mathcal{O}_\mathcal{K})$ to the various one–parameter subgroups $\psi : G_m \to T$. On a
weight vector $v \in M$ of weight μ, such a subgroup acts by

$$\psi(t) v = t^{(\mu, \nu)} v \qquad (t \in k^*),$$

where ν is the integral weight corresponding to ψ (that is $\nu = (d\psi)(1)$, cf. 3.16).

This shows that in our notation,

$$R_M(t,\nu) = \overset{*}{\psi}(ch(M)) \,,$$

that is Joseph's one variable Poincaré series is just the restriction of the formal

character of M to a one–parameter subgroup. As discussed in detail in 3.16, this

one–variable series is obtained from the multi–variable power series

$ch(M) = \underset{\mu}{\Sigma} (\dim M_\mu)e^\mu$ by "specializing" the variables $t_i = e^{\alpha_i}$ (i = 1,...,ℓ) all to a

single variable t via $t_i \longmapsto t^{(\alpha_i,\nu)}$. This specialization may also be interpreted as

an "evaluation" of ch(M), considered as a function on $t = \overset{*}{t}$ putting $e^\tau = t$. So

we consider

$$\overset{*}{\psi}(ch(M)) = ch(M)(\tau\nu) = R_M(t,\nu) \qquad (3)$$

as the (formal) evaluation of ch(M) on the Lie algebra of our one–parameter

subgroup. Similarly, we get

$$\overset{*}{\psi}(\Delta(\overset{*}{u})ch(M)) = \underset{\alpha>0}{\Pi} (1-t^{(\alpha,\nu)})R_M(t,\nu) \qquad (4)$$

which we denote by $P_M(t,\nu)$. We know from §3, that this is a Laurent polynomial

$P_M(t,\nu) \in k[t,t^{-1}]$; in fact, it is the class determined by the restriction of M (with

respect to ψ) in the equivariant K–group

$$K_{\psi(G_m)}(u) \simeq R(G_m) \simeq k[t,t^{-1}] \,.$$

Now let us assume again that ν is regular (all $(\alpha,\nu) > 0$, for all $\alpha > 0$). Then we

know that the order of vanishing of $P_M(t,\nu)$ at $t = 1$ is $d = \text{codim}_u C$ (cf. 3.18),

so let us write

$$P_M(t,\nu) = (t-1)^d \tilde{P}_M(t,\nu) \,,$$

where now $\tilde{P}_M(t,\nu)$ is a Laurent polynomial such that $\tilde{P}_M(1,\nu) \neq 0$. From the

resulting expression

$$R_M(t,\nu) = \tilde{P}_M(t,\nu) \frac{(t-1)^d}{\underset{\alpha>0}{\Pi} (1-t^{(\alpha,\nu)})} \,, \quad (\tilde{P}_M(1,\nu) \neq 0) \qquad (5)$$

for Joseph's Poincaré series, one may now deduce some information about its Hilbert–Samuel function (by the methods used in [AC] VIII, §4. cf. also [Sm], or [J1], or our 3.17): It has degree $n-d = \dim C$ (cf. 3.18), and leading coefficient $\tilde{P}_M(1,\nu)/(n-d)!$. Comparing this with Joseph's notation (see 1), we get

$$\tilde{p}_M(1,\nu) = p_M(\nu) = p_C(\nu) .$$

Combining this with (5) and (4), we obtain that

$$\psi^*(\Delta(u^*)ch(M)) = p_C(\nu)(t-1)^d + \ldots = p_C(\nu) \cdot \tau^d + \ldots$$

up to higher terms in $t-1$ resp. in τ, so the homogeneous term of degree d must map to

$$\psi^*([\Delta(u^*)ch(M)]^d) = p_C(\nu)\tau^d . \tag{6}$$

On the other hand, ψ^* means evaluation at $\tau\nu$ by (3) above, and $[\Delta(u^*)ch(M)]^d$ is our "character polynomial" q_M^T by corollary 3.9 (which has degree d by theorem 3.10). Hence equation (6) reads as follows: $q_M^T(-\mathfrak{t}) = p_C(\nu)\tau^d$. But $q_M^T(\tau\nu) = q_M^T(\nu)\tau^d$ by homogeneity, and so we conclude that our polynomial q_M^T coincides with Joseph's p_C. Finally, we have also $q_M^T = Q_G(K)$, as was already explained in 4.5.

4.9. Generalization to the case of sheaves.

Let us extend the considerations of the previous section to the case of an arbitrary G–equivariant conical coherent sheaf $E \neq 0$ on T^*X. Let $F = j^*E$ denote the corresponding B–equivariant coherent sheaf on the fibre u of T^*X, obtained by restriction. As before, we let M denote the T–equivariant $S(u^*)$ module $M = \Gamma(u,E) = \Gamma(u,F)$. Generalizing proposition 4.8, we may now state:

Proposition 1: *Considered as polynomials on* $t = t^*$, *our equivariant characteristic class* $Q_G(E) \ (= q_F^T)$, *and Joseph's "characteristic polynomial"* $p_{S(M)}$, *as defined in [J1], 5.5, coincide.*

Let us denote the supporting cycles for E resp. F (cf. 1.4) by

$$\text{supp}(E) = \sum_i m_i(E)[\mathcal{U}_i] \qquad \text{resp.}$$

$$\text{supp}(F) = \sum_i m_i(F)[\mathcal{V}_i]$$

Then the irreducible components \mathcal{U}_i resp. \mathcal{V}_i are clearly G–resp. B–stable closed cone bundles resp. cones, and by choosing the correct numbering, each \mathcal{U}_i will be a homogeneous cone bundle with fibre \mathcal{V}_i, that is $\mathcal{U}_i = G \times^B \mathcal{V}_i$ for all i. Moreover, the multiplicities and codimensions will coincide:

$$m_i(E) = m_i(F) \text{ for all i,}$$

and

$$\text{codim}_{T^*X} \text{supp}(E) = \text{codim}_u \text{supp}(F) =: d .$$

Proposition 2: With notations as above,

(a)
$$Q_G(E) = \sum_{\text{codim } \mathcal{U}_i = d} m_i(E) Q_G(\mathcal{U}_i)$$

b)
$$q_F^T = \sum_{\text{codim } \mathcal{U}_i = d} m_i(F) q_{\mathcal{V}_i}^T .$$

Proof of proposition 2: As explained in 4.5, we have $Q_G(E) = q_F^T$, and $Q_G(\mathcal{U}_i) = q_{\mathcal{V}_i}^T$, that is statements a) and b) coincide term by term. In the Grothendieck group $K_T(u)$, we have up to higher degree terms:

$$[F] = \sum_i m_i(F)[\mathcal{O}_{\mathcal{V}_i}] .$$

By theorem 3.10, we know (since E and hence F and M are $\neq 0$) that the γ–degree of $[F]$ is d, and that of $[\mathcal{V}_i]$ is codim \mathcal{V}_i, so in $\text{gr}_d K_T(u)$ we get

$$[F] \equiv \sum_{\text{codim } \mathcal{V}_i = d} m_i(F)[\mathcal{O}_{\mathcal{V}_i}] \mod K_T^{d+1}(u) .$$

By definition of our character polynomials, this means exactly

$$q_F^T = \sum_{\text{codim } \mathcal{V}_i = d} m_i(F) q_{\mathcal{V}_i}^T \qquad\qquad \text{Q.e.d.}$$

Corollary: *The equivariant characteristic class of* **E** *is completely determined by its* (G–*equivariant!*) *characteristic cycle.*

So far, we have established an equivariant version of proposition 1.4. Let us now make the link to Joseph's work, i.e., let us *prove proposition* 1 above. Joseph's definition of a character polynomial p_M, as recalled in 4.8, generalizes in an obvious way to our present module $M = \Gamma(u, E) = \Gamma(u, F)$, and then we get the coincidence $Q_G(E) = q_F^T = p_M$ by the same proof as before. Note, however, that Joseph in loc.cit. does not explicitly use this obvious extension of his notation. Instead, he introduces the new notation $p_{S(M)}$ (see loc.cit., 5.5). In our terminology, his $S(M)$ is the top–dimensional part of the supporting cycle of M (that is of F):

$$S(M) = \sum_{\text{codim } \mathcal{V}_i = d} m_i(F)[\mathcal{V}_i] .$$

Then he *defines*

$$p_{S(M)} := \sum_{\text{codim } \mathcal{V}_i = d} m_i(F) p_{\mathcal{V}_i} .$$

But now it suffices to observe that $p_{\mathcal{V}_i} = q_{\mathcal{V}_i}^T$ by 4.8, and so $p_{S(M)} = p_M$ by proposition 2b) above. This completes the proof of proposition 1. Q.e.d.

Remark: From now on, we shall normally no longer care much about notational distinctions between Joseph's characteristic polynomials p and our character polynomials q.

4.10 Equivariance under a Levi subgroup.

Now let G' be a connected reductive closed subgroup of G containing T. We denote $W' \subset W$ its Weyl group, by U' the maximal unipotent subgroup $U' = G' \cap U$ and by U'_- the "opposite unipotent subgroup $U'_- = G' \cap U_-$. Let

us now assume that our coherent sheaf \mathbf{F} on \mathfrak{u}, viewed as a coherent sheaf on \mathfrak{g}, is not only $T-$, but even G'—equivariant. Then $M = \Gamma(\mathfrak{u},\mathbf{F})$ is a direct sum of finite–dimensional simple G'—modules, and so its formal character is of the form

$$\text{ch}(M) = \sum_{\lambda \in \Lambda} \text{mtp}(L'(\lambda),M)\text{ch}(L'(\lambda)) , \qquad (1)$$

where $L'(\lambda)$ denotes the simple G'—module of highest weight λ. The λ—weight space of $L'(\lambda)$ equals the space of U'—invariants,

$$L'(\lambda)_\lambda = L'(\lambda)^{U'} ,$$

and since it is one dimensional, we may also write

$$\text{mtp}(L'(\lambda),M) = \dim M_\lambda^{U'-} \qquad (2)$$

for the multiplicity of $L'(\lambda)$ in M.

Lemma: $\text{ch}(M) = \Delta(\mathfrak{u}_-'^*)^{-1} e^{-\rho'} (\sum_{w \in W'} \det(w)w)(e^{\rho'} \text{ch}(M^{U'-}))$. Here \mathfrak{u}' is the Lie algebra of U', ρ' is half the sum of roots occurring in \mathfrak{u}_-', and $M^{U'-}$ is the space of U_-' invariants in M.

Proof: By Weyl's character formula ([Hu], 24.3),

$$\text{ch}(L'(\lambda)) = (\sum_{w \in W'} \det(w)e^{w(\lambda+\rho')})/(\sum_{w \in W'} \det(w)e^{w\rho'}) \qquad (3)$$

By the denominator formula (cf. loc.cit.), we have

$$\sum_{w \in W'} \det(w)e^{w\rho'} = \prod_{\alpha>0} (e^{\alpha/2}-e^{-\alpha/2}) = \Delta(\mathfrak{u}_-'^*)e^{\rho'} \qquad (4)$$

(notation 3.3). Inserting (3), (4), (2) into (1) above, we obtain

$$\text{ch}(M) = \sum_\lambda (\dim M_\lambda^{U'-})(\sum_{w \in W'} \det(w)e^{w(\lambda+\rho')})e^{-\rho'} \Delta(\mathfrak{u}_-'^*)^{-1} .$$

Further computation gives then

$$e^{\rho'} \Delta(u_-^{'*}) ch(M) = \sum_{w \in W'} \sum_{\lambda} (\dim M_\lambda^{U'_-}) \det(w) w(e^{\lambda + \rho'})$$

$$= \sum_{w \in W'} \det(w) w \sum_{\lambda} (\dim M^{U'_-}) e^{\lambda} e^{\rho'})$$

$$= (\sum_{w \in W'} \det(w) w)(e^{\rho'} ch(M^{U'_-})),$$

which is the formula claimed in the lemma. Q.e.d.

Proposition: *If* F *is* G'*—equivariant, then its character polynomial* $q_F = q_M$ *is* W' *anti—invariant* (i.e., $sq_M = -q_M$ *for each simple reflection* s *in* W').

Proof: Since q_M is the lowest degree term of $\Delta(u^*) ch(M)$, and since the lowest degree term of $\Delta(u^*)$ equals $\prod_{\alpha > 0} \alpha$, which is clearly anti—invariant, it is enough to show that $ch(M)$ itself is W'—invariant. This follows from the formula in the above lemma, or more directly from (1) and the W'—invariance of $ch(L'(\lambda))$.

4.11 Multiple cross—section of a unipotent action.

With assumptions and notations as in 4.10, let us now study further the case $F = \mathcal{O}_V$, where $V \subset u$ is a G'—stable closed irreducible subvariety. The ring of U'_- invariants $A := \mathcal{O}(V)^{U'_-}$ is known to be finitely generated by a theorem of Hadžiev ([Ha], cf. also [Kr]), and so we can define an affine variety $V/U'_- := \operatorname{Spec} A$, called the quotient of V by U'_-. It is equipped with the canonical morphism π : $V \to V/U'$ provided by the inclusion $A \subset \mathcal{O}(V)$. Generically, this morphism is a "good quotient map" for the group action, that is an (affine) fibration with a single U'—orbit as a fibre, see e.g. [BGR], Satz 16.6; more precisely, for a suitable highest weight vector $0 \neq a \in A$, the localization $\mathcal{O}(V)_a$ is a polynomial ring over A_a, and

$$\mathcal{O}(V)_a = A_a \otimes \mathcal{O}[V]$$

with V an affine homogeneous U'—space, U'_- acting via its action on the second

factor (cf. loc.cit.). In particular, dim $V/U'_- = d(A) = \dim V - \dim V$, that is the \mathcal{GK}–dimension of the U'_-–invariants is given by the dimension of the variety, minus the "generic orbit dimension."

Let us now make the assumption, for simplicity, that this latter dimension is as big as possible, that is that V contains a free U'_-–orbit [hence a generic subset of them, cf. loc.cit., or [BK]). Following [J1], 2.6, and using our notation ψ, ν from 4.8 for a one–parameter group, we consider the Poincaré series

$$R_A(t,\nu) := \sum_{\mu \in \Lambda} (\dim A_\mu) t^{(\mu,\nu)} = \psi^*(\mathrm{ch}(A))$$

and *define* a function $r_A(\nu)$ and a polynomial on $t^* (= t)$

$$p_A(\nu) := r_A(\nu) \prod_{\alpha > 0} \alpha(\nu)$$

in a manner completely analogous to the procedure of 4.8, which defined $R_M(t,\nu) = \psi^*(\mathrm{ch}(M))$, $r_M(\nu)$, and $p_M(\nu)$ for $M = \mathcal{O}(V)$. We shall also use the notations $p_A = p_{V/U'_-}$, and $p_M = p_V$. We note that these polynomials coincide with $q_A = q_{V/U'_-}$, as defined in 3.19.

Lemma A: *If U'_- acts generically freely on the G'–stable subvariety $V \subset u$ as above, then:*

$$p_V = \prod_\beta \left(\tfrac{1}{\beta}\right) \sum_{w \in W'} {}^w p_{V/U'_-},$$

where the product extends over all roots β occurring in U'_-.

Proof: By restriction to a one parameter subgroup, lemma 4.10 gives:

$$R_M(t,\nu) = \prod_\beta \frac{1}{(1-t^{-(\beta,\nu)})} e^{-\rho'} \sum_{w \in W'} \det(w) w \cdot e^{\rho'} R_A(t,\nu)$$

$$= \prod_\beta (1-t^{-(\beta,\nu)})^{-1} \sum_{w \in W'} \det(w) w \, R_A(t,\nu) + \cdots$$

where \cdots are power series of higher lowest degree term, which may be neglected in

computing r_V and $r_{V/U_-'}$. Now by taking lowest order terms, as in the special case considered in [J1], 2.6, we get

$$r_M(\nu) = \underset{\beta}{\Pi'} (\beta,\nu)^{-1} \underset{w\in W}{\Sigma'} \det(w) w \, r_A(\nu) ,$$

and so multiplication by the product of all positive roots, which is W' anti–invariant, gives the lemma.

Now we consider a B–stable "multiple cross section" C for the U_-'–action on V. By this, we mean a closed (T–stable) irreducible subvariety $C \subset V$ of dimension $\dim C = \dim V/U_-' = \dim V - \dim U_-'$, such that $\overline{U_-C} = V$.

By the last assumption, the restriction of functions on V to C embeds $A = \mathcal{O}(V)^{U_-'}$ into $\mathcal{O}(C)$ as a subring, and by the first assumption, $d(A) = d(\mathcal{O}(C))$, so that the restriction of the quotient map $V \to V/U_-'$ gives a generically *finite map* $C \to V/U_-'$. We denote $d(C,V/U_-')$ the degree of this map, or in other words the degree of the field extension

$$d(C,V/U_-') = [\text{Fract}(\mathcal{O}(C)) : \text{Fract } A] .$$

Note that *this is the number of times that a generic orbit in V meets C.*

Lemma B:
$$\underset{w\in W}{\Sigma'} w \cdot p_C = d(C,V/U_-') \cdot \underset{w\in W}{\Sigma'} w \cdot p_{V/U_-'} .$$

We introduce the G'–variety $Y := G' \overset{,B'}{\times} C$ and the G'–morphism $q : Y \longrightarrow V$ defined by $q(g,\xi) = g \cdot \xi$. q is the composition of the closed immersion $G' \overset{,B'}{\times} C \hookrightarrow G' \overset{,B'}{\times} V$, of the isomorphism $\varphi : G' \overset{,B'}{\times} V \overset{\sim}{\longrightarrow} (G'/B') \times V$ defined by $\varphi(g,\xi) = (g,g^{-1}\xi)$ and of the projection $(G'/B') \times V \longrightarrow V$. Since each of these three morphisms is proper, q is proper too. Hence by Grauert's theorem, $q_*(\mathcal{O}_Y)$ is a coherent \mathcal{O}_V–module; since V is affine, $\mathcal{O}(Y) = \Gamma(V,q_*(\mathcal{O}_Y))$ is a finitely generated $\mathcal{O}(V)$–module. The degree z of q is the same as the degree

of the map $U'_- \times \mathcal{V} \longrightarrow \mathcal{V}$ given by the U'_-–action since this last map is the composition of an open dimension $U'_- \times \mathcal{V} \longrightarrow Y$ and of $q : Y \longrightarrow \mathcal{V}$; it's then clear that $z = d(C, \mathcal{V}/U'_-)$. Hence, from Proposition 3.14, we obtain:

$$p_Y := p(\mathcal{O}(Y)) = z \cdot p_{\mathcal{V}}.$$

We will also need to consider the higher cohomology groups $H^i(Y, \mathcal{O}_Y)$. These are G'–modules, equal to $\Gamma(\mathcal{V}, R^i q_*(\mathcal{O}_Y))$. Again by Grauert's theorem, each $R^i q_*(\mathcal{O}_Y)$ is a coherent sheaf. Since q is generically finite, the support of $R^i q_*(\mathcal{O}_Y)$ has codimension ≥ 1 in \mathcal{V}. Hence the character polynomial of these G'–equivariant coherent sheaves has degree bigger than the degree of p_Y (from theorem 3.10). If we introduce the virtual G'–equivariant coherent sheaf

$$\chi(Y) := \sum_{i \geq 0} (-1)^i [R^i q_*(\mathcal{O}_Y)],$$ we can still define its character polynomial, $p(\chi(Y))$, and we find that $p(\chi(Y)) = p_Y$. This means that p_Y is the lowest order term of $\Delta(\mu^*) \sum_{i \geq 0} (-1)^i ch(H^i(Y, \mathcal{O}_Y))$.

Now we analyze $\sum_{i \geq 0} (-1)^i H^i(Y, \mathcal{O}_Y)$ from the G'–equivariant projection p: $Y \longrightarrow G'/B'$. This is an affine morphism, with fiber C. Hence $H^i(Y, \mathcal{O}_Y) = H^i(G'/B', p_*(\mathcal{O}_Y))$. Now $p_*(\mathcal{O}_Y)$ is a G'–equivariant locally free sheaf on G'/B' (of infinite rank), corresponding to the (locally finite) B'–module $\mathcal{O}(C)$. As a B'–module, $\mathcal{O}(C)$ may be filtered by finite–dimensional B'–submodules, with one–dimensional subquotients k_λ. Now $ch(\chi(G'/B', \mathcal{L}(\lambda))$ is given by the Borel–Weil–Bott theorem. Precisely, $ch(\chi(G'/B',$

$$\mathcal{L}(\lambda)) = e^{-\rho'} \cdot \Delta(\mu_-^*)(\sum_{w \in W'} \det(w)w)(e^{\rho'} e^\lambda) \text{ (with the notations of 4.10)}.$$

Summing up over all weights λ occurring in $\mathcal{O}(C)$, we get:

$$ch(\chi(Y)) = ch(\chi(G'/B', p_*(\mathcal{O}_Y)) = e^{-\rho'} \cdot \Delta(u_-^*)(\sum_{w \in W'} \det(w)w)(e^{\rho'} \cdot ch(\mathcal{O}(C)))$$

Taking the character polynomials as in 4.10, we obtain

$$p(\chi(Y)) = \prod_\beta{}' (\tfrac{1}{\beta})(\sum_{w \in W'} w)(p_C)$$

(again the product is over the roots β occurring in u'). Now using Lemma A, we obtain

$$\prod_\beta{}'(\tfrac{1}{\beta})(\sum_{w\in W'}w)(p_C) = z \cdot \prod_\beta{}'(\tfrac{1}{\beta})(\sum_{w\in W'}w)(p_{V/U_-'})$$

One obtains the lemma by multiplication with $\prod_\beta{}'\beta$.

4.12 For example SL_2–equivariance.

Let us specialize the previous discussions to the case where G' is of type A_1, and $U_-' = U_{-\alpha}$ for a simple root α of G. (Here we denote U_β the root subgroup of G, for any β). If $C \subset u$ is any closed irreducible subvariety which is stable under $B_\alpha = TU_\alpha$, but not under $G_\alpha = G'$, then its G_α–saturation $V := G_\alpha C$ will be a closed (!) G_α–subvariety (use [St2], p. 68, lemma 2) satisfying our assumptions in 4.11, provided that $G_\alpha C \subset u$, and we obtain as special cases of 4.10, 4.11:

Lemma (cf. [J1], 2.5–2.7)[1]: *Let* C *be an irreducible closed* B_α–*stable subvariety of* u, *and assume* $G_\alpha C \subset u$. *Then*

$$-\frac{1}{\alpha}(s_\alpha+1)\,p_C = z \cdot p_{G_\alpha C}\,,$$

where $z = 0$ *if* C *is* G_α–*stable, and* $z = d(C,V/U_{-\alpha}) > 0$ *otherwise.*

4.13 Completing the proof of theorem 4.7.2.

We are now ready to complete the proof of our main results announced in 4.7. The argument below is due to Joseph [J1], 3.1, and is repeated for convenience of the reader in our present framework.

[1]In comparing with [J1], note that there are sign errors in propositions 2.6 and 2.7, and in lemma 2.9 of [J1], which happen to cancel each other in the calculation in [J1], 3.1.

Let $s \in W$ be a simple reflection, α the corresponding simple root and $P_s = B \cup sBs^{-1}$ the corresponding minimal parabolic subgroup of G. For all orbital cone bundles $\mathcal{K}_i = G \times^B C_i$, we wish to compute $sQ_G(\mathcal{K}_i)$ as a linear combination of $Q_G(\mathcal{K}_j)$'s, where $i,j = 1,\ldots,r$. By 4.6 resp. 4.8, we may equivalently work in terms of the cones C_j and their character polynomials $q_T(C_j)$ resp. characteristic polynomials p_{C_j}. If C_i should be P_s–stable, then

$$sp_{C_i} = -p_{C_i}$$

by 4.12, and we are done.

So let us assume from now on that C_i is *not* P_s–stable. Let u_s denote the hyperplane in u orthogonal to the root space for $-\alpha$. This hyperplane is also the nilradical of the Lie algebra of P_s. In particular, u_s is P_s–stable. We conclude that our cone C_i is not contained in this hyperplane, because otherwise also $P_s C_i$ would be contained in $u_s \subset u$, and in $GC_i = \overline{\mathcal{O}}_u$, contradicting the fact that C_i is a maximal irreducible subset of $\overline{\mathcal{O}}_u \cap u$. Therefore, u_s intersects C_i in codimension one; more precisely, by lemma 3.11, the corresponding character (istic) polynomials are related by

$$p_M = \alpha p_{C_i} \, , \tag{1}$$

where the module M describes the intersection (including multiplicities),

$$M := \Gamma(C_i \cap u_s, \mathcal{O}_{C_i}) = \mathcal{O}(C_i)/x_{-\alpha} \mathcal{O}(C_i) \, ,$$

where $x_{-\alpha}$ is a root vector for $-\alpha$. Let us denote its supporting cycle as in 1.4, i.e.

$$\text{supp}(M) = \sum_{\mathcal{V}} m_{\mathcal{V}}(M) \, [\mathcal{V}] \, . \tag{2}$$

Here the irreducible components \mathcal{V} of $C_i \cap u_s = \text{supp}(M)$ are all of codimension one in C_i, by Krull's Hauptidealsatz. Now we claim that for each such component \mathcal{V}, the P_s–saturation $P_s \mathcal{V}$ is either \mathcal{V}, or else one of the other C_j's, for some $j = j(s,\mathcal{V}) \neq i$. In fact, since each \mathcal{V} is B–stable, it follows that either it is P_s–stable, or else its P_s–saturation has dimension

$$\dim P_s \mathcal{V} = 1 + \dim \mathcal{V} = \dim C_i \ .$$

But since $P_s \mathcal{V}$ is irreducible and contained in $\overline{\mathcal{O}_u} \cap u$ (in fact even in u_s), which is equidimensional, it follows that then $P_s \mathcal{V} = C_j$ for some $j = j(s, \mathcal{V}) \neq i$. Now we apply 4.12 to conclude that

$$-\frac{1}{\alpha} (1+s) \, p_{\mathcal{V}} = z p_{C_j} \tag{3}$$

for $j = j(s, \mathcal{V})$ and some integer

$$z = z(s, \mathcal{V}) = d(\mathcal{V}, C_j / U_{-\alpha}) \ . \tag{4}$$

Note that (3) holds for all \mathcal{V}, if we just put $z(s, \mathcal{V}) = 0$ in case $P_s \mathcal{V} = \mathcal{V}$ (use 4.11).

Now we have

$$p_{C_i} = \alpha^{-1} \, p_M = \alpha^{-1} \sum_{\mathcal{V}} m_{\mathcal{V}}(M) \, p_{\mathcal{V}}$$

by (1), (2), and 4.9, and so we compute, using (3):

$$(s-1) p_{C_i} = -\alpha^{-1}(s+1) p_M = -\alpha^{-1} \sum_{\mathcal{V}} m_{\mathcal{V}}(M)(s+1) p_{\mathcal{V}}$$

$$= \sum_{\mathcal{V}} m_{\mathcal{V}}(M) z(s, \mathcal{V}) p_{C_{j(s, \mathcal{V})}} \ . \qquad \text{Q.e.d.}$$

4.14 Reproving Hotta's transformation formula.

The proof yields the following more precise formula. A subset of $T^* X$ is called s–*vertical* (cf. Hotta's terminology [Ho]), if it is a union of projective lines of type s, i.e., projecting onto a conjugate of the line P_s / B in $X = G/B$.

Corollary: If \mathcal{K}_i is s–vertical, then $s Q_G(\mathcal{K}_i) = -Q_G(\mathcal{K}_i)$, and otherwise

$$s Q_G(\mathcal{K}_i) = Q_G(\mathcal{K}_i) + \sum_j n^s_{ij} \, Q_G(\mathcal{K}_j) \ , \tag{4}$$

where $n^s_{ij} = 0$ unless \mathcal{K}_j is s–vertical and meets \mathcal{K}_i in codimension 1, in which case

$$n^s_{ij} = \sum_{\mathcal{W}} z_{\mathcal{W}} m_{\mathcal{W}} \geq 0 \ .$$

Here the summation is over all irreducible components \mathcal{W} of $\mathcal{K}_i \cap \mathcal{K}_j$ (necessarily of codimension one), $m_{\mathcal{W}}$ is the multiplicity introduced in 4.15, and $z_{\mathcal{W}}$ is the non–negative integer $z(s, \mathcal{W} \cap u)$ defined in 4.13.

Remark: (1) By applying the canonical map $K_G(X) \to K(X) \cong H^*(X)$, which forgets the G–action, to (4), we reobtain Hotta's formula in the version of our theorem 1.15.

(2) According to Hotta [Ho], $m_{\mathcal{W}}$ is the interaction multiplicity of \mathcal{K}_i and \mathcal{K}_j at \mathcal{W}.

4.15 On explicit computations of our characteristic classes.

Let us conclude this chapter with a few remarks and examples concerning the explicit computation of our characteristic classes $Q(\mathcal{K}_i)$ in $H^*(X)$ introduced in chapter 1. The first point to be made here is that it is more convenient to perform the computation on the level of equivariant K–theory, using the fact that the cone bundle under consideration is a G–equivariant one. Of course, to know $Q(\mathcal{K}_i)$, or to know $Q_G(\mathcal{K}_i)$, amounts to the same, in view of 4.6 and 4.7 (theorem 1). However, actual computations tend to be much more pleasant in $K_G(X) \cong R(T)$, a unique factorization domain, than in $H^*(X)$, which has lots of nilpotent elements.

The second point is that the interpretation of $Q_G(\mathcal{K}_i)$ as a character polynomial $q_T(C_i)$ by 4.5 is helpful for calculations. (Here C_i denotes the fibre of \mathcal{K}_i.) Let us give an example.

Proposition: *Suppose* C_i *is a complete intersection of codimension* d *in* u, *defined by a regular sequence of equations* $f_1, \ldots, f_d \subset \mathcal{O}(C_i)$ *which are* T–*semi–invariant, of weights* $-\mu_1, \ldots, -\mu_d$. *Then*

$$Q_G(\mathcal{K}_i) = P_{C_i} = \mu_1\mu_2\cdots\mu_d \,.$$

Proof: This follows by repeated application of lemma 3.12. Q.e.d.

For example, if f_1,\ldots,f_d are root vectors $x_{-\alpha_1},\ldots,x_{-\alpha_d}$, then

$Q_G(\mathcal{K}_i) = \alpha_1\alpha_2\cdots\alpha_d$, and we recover the example treated in 1.6. In this example, all equations are linear, and the cone bundle \mathcal{K}_i is actually a vector bundle, so all computations are very easy. Let us therefore conclude with a less trivial case, where \mathcal{K}_i is not a vector bundle, and which is also covered by the proposition above.

4.16 Example.

Take $G = SL_n$. For $n \leq 3$, all orbital cone bundles are vector bundles, and the computation of their characteristic classes are covered by 1.6. For $n = 4$, there occurs the following orbital cone C, which is not linear, but quadratic: It consists of all block triangular matrices of rank $\leq r$ of the form

$$\begin{bmatrix} 0 & A \\ 0 & 0 \end{bmatrix} \;,\; \text{where } A = \begin{bmatrix} a_{1,p+1} & \cdots & a_{1,n} \\ \vdots & & \vdots \\ a_{p,p+1} & \cdots & a_{p,n} \end{bmatrix} \tag{*}$$

is a p by q matrix of rank $\leq r$. This cone is denoted $C(r,p,q)$. Then $C(0,p,q) \subset C(1,p,q) \subset \ldots \subset C(p,p,q)$ is a chain of orbital cones, and these are all the orbital cones which are contained in the vector space of all matrices of this form (*). Now we take the particular case $p = q$, and $r = q-1$: The cone $C(p-1,p,p)$ is again a complete intersection, given by the linear equations $a_{ij} = 0$ $(1 \leq i,j \leq p,$ or $p \leq i,j \leq n)$, and the single non-linear equation $\det A = 0$, of weight

$$\mu = \alpha_1 + 2\alpha_2 + 3\alpha_3 + \ldots + p\alpha_p + (p-1)\alpha_{p+1} + \ldots + \alpha_n$$
$$= p\alpha_p + \sum_{i-1}^{p-1} i(\alpha_i + \alpha_{n-i}) \,.$$

So again by 4.14, we conclude that

$$P_{C(p-1,p,p)} = (\alpha_{1,n} + \alpha_{2,n-1} + \ldots + \alpha_{p,p+1}) \prod_{1 \leq i < j \leq p} \alpha_{ij} \prod_{p \leq i < j \leq n} \alpha_{ij} \,.$$

For other values r,p,q, the cone C(r,p,q) need no longer be a complete intersection, and the computation of characteristic classes becomes more delicate.

4.17 Remark. A possible geometric generalization of proposition 4.15.

We *conjecture* (along with W. Fulton) the following generalization of the Whitney sum formula for Chern classes of vector bundles ([Fu], p. 51) to Segre classes of cone bundles. Suppose we have a diagram of cone bundles

$$\mathcal{K} \hookrightarrow E \xrightarrow{\oplus f_i} \bigoplus_{i=1}^{m} \ell_i$$

where the vector bundle E is mapped by a fibre preserving algebraic but not necessarily linear map to a direct sum of line bundles, and the local complete intersection cone bundle \mathcal{K} of codimension m in E is defined scheme theoretically by the vanishing of $\oplus f_i$. Then

$$s(E) = s(\mathcal{K})\, s(\oplus \ell_i) \, ,$$

or, equivalently, the Segre class of \mathcal{K} may be computed from the Chern classes of $E, \ell_1, ..., \ell_m$, as

$$s(\mathcal{K}) = c(\ell_1)...c(\ell_m)/c(E) \, .$$

Notice that a special case of this is implied by proposition 4.15. This illustrates the power of the methods of equivariant K–theory in algebraic geometry.

§5. Characteristic Classes and Primitive Ideals

In this chapter, G is again a semisimple group with Lie algebra \mathfrak{g}, and we use the notations introduced in 1.5. As in chapter 3 and 4, we denote by Λ the lattice in t^* of integral weights (3.1) of our maximal torus $T \subset G$, by $\Omega \subset \Lambda$ the "dominant integral weights" (3.16) with respect to the ordering fixed by our choice of a Borel subgroup $B \supset T$, and by $\rho \in \Omega$ half the sum of weights in \mathfrak{b}. We furthermore denote by $\mathcal{U}(\mathfrak{g})$ the enveloping algebra of \mathfrak{g}, that is the ring of differential operators on G invariant under right translations. Our purpose is to study \mathfrak{g}–modules, that is to say $\mathcal{U}(\mathfrak{g})$–modules. In particular, we are interested in the annihilators in $\mathcal{U}(\mathfrak{g})$ of simple \mathfrak{g}–modules, called primitive ideals. We denote by $L(\lambda)$ the simple \mathfrak{g}–module of highest weight λ (which is defined as the unique simple quotient of the universal (or Verma–) module $M(\lambda) = \mathcal{U}(\mathfrak{g}) \otimes_{\mathcal{U}(\mathfrak{b})} k_\lambda$, where k_λ is a one–dimensional \mathfrak{b}–module of weight λ). Then the center of $\mathcal{U}(\mathfrak{g})$ is a polynomial ring in dim (T) variables (Harish–Chandra, Chevalley); this center acts by a character on $L(\lambda)$ which is denoted χ_λ; we note that by Harish–Chandra's theorem, $\chi_\lambda = \chi_\mu$ if and only if $\mu = w.\lambda$, for some Weyl group element $w \in W$, where the "shifted Weyl group action" $w.\lambda := w(\lambda+\rho)-\rho$ is used. Finally, it will be convenient to identify \mathfrak{g}^* with \mathfrak{g}, and t^* with t by means of the Killing form. We apply analogous notations to the group $G \times G$, so for instance $(\lambda,\mu) \in t^* \times t^*$ defines a central character $\chi_{(\lambda,\mu)}$ of $\mathcal{U}(\mathfrak{g} \times \mathfrak{g})$ etc..

5.1. Characteristic class attached to a \mathfrak{g}–module.

Let M be a finitely generated \mathfrak{g} module, with central character χ_λ, given by the dominant integral weight λ. Let $\mathcal{M} = \mathcal{D}_X^\lambda \otimes_{\mathcal{U}(\mathfrak{g})} M$ denote its localization as a coherent \mathcal{D}_X^λ module on X [BeBe], where \mathcal{D}_X^λ is the sheaf of differential operators on X with coefficients in the line bundle $\mathcal{L}(\lambda)$ given by λ. The notation we use is slightly different from that of [BeBe]. Choose a good filtration on

M, and consider the associated graded sheaf gr\mathcal{M} as a coherent sheaf on T^*X. Then we define the *characteristic variety* of M by

$$Ch\,(M) := Ch\,(\mathcal{M}) = \text{supp gr }\mathcal{M}\,,$$

the *characteristic cycle* of M by (notation 1.4)

$$Ch\,(M) := Ch\,(\mathcal{M}) := \text{supp (gr }\mathcal{M})\,,$$

and the *characteristic class* associated to M by (notation 1.3)

$$P(M) := Q(\text{gr }\mathcal{M})\,.$$

This is a cohomology class in $H^{2d}(X)$, where d is the codimension of $Ch(M)$ in T^*X; if convenient, we also consider $P(M)$ as a harmonic polynomial, homogeneous of degree d on t (convention 1.6). Let us point out that the above notions are *well–defined*, i.e. independent of the choice of a good filtration. For $P(M)$, this is proved the same way as for $Ch(\mathcal{M})$ (or a special case, $G = 1$, of Lemma 2.12). In fact, even more is true:

Proposition: *The characteristic class of a module is entirely determined by its characteristic cycle. More precisely, if* $V_1,...,V_r$ *are the irreducible components of* $Ch(M)$ *of minimal codimension* d, *with multiplicities* $m_1,...,m_r$ *in* $Ch\,(M)$, *then* (notation 1.4)

$$P(M) = m_1\,Q(V_1) + ... + m_r\,Q(V_r)\,.$$

This is true by proposition 1.4.

5.2 Translation invariance.

Let λ,μ be dominant integral weights. Let $E(\mu-\lambda)$ denote the finite dimensional simple \mathfrak{g}–module with extremal weight $\mu-\lambda$. Let M be a \mathfrak{g}–module with central character χ_λ. Then $T^\mu_\lambda M$ is *defined* as the direct summand of central character χ_μ of $M \otimes E(\mu-\lambda)$. The *"translation functor"* T^μ_λ is then an

equivalence of the categories of finitely generated \mathfrak{g}–modules with central character χ_λ resp. χ_μ ([BeGe], [Ja]).

Lemma: $\mathit{Ch}(T_\lambda^\mu M) = \mathit{Ch}(M)$, and $P(T_\lambda^\mu M) = P(M)$.

Proof: The second statement follows from the first one by proposition 5.1. Let \mathcal{M} resp. $T_\lambda^\mu \mathcal{M}$ denote the \mathcal{D}_X^λ–resp. \mathcal{D}_X^μ–module corresponding to M resp. $T_\lambda^\mu M$. Then the functor T_λ^μ thus defined is equivalent to the "*geometric translation functor*" $\mathcal{M} \longmapsto \mathcal{O}(\mu{-}\lambda) \otimes_{\mathcal{O}_X} \mathcal{M}$, where $\mathcal{O}(\mu{-}\lambda)$ denotes the invertible sheaf corresponding to the line bundle $\mathcal{L}(\mu{-}\lambda)$. Since $\mathcal{O}(\mu{-}\lambda)$ is locally isomorphic to \mathcal{O}_X, it follows that this functor does not change the characteristic cycle. Q.e.d.

Remark: By the lemma, all of our results stated in [BB3] for the trivial central character (case $\lambda = 0$) only, extend to the case of an arbitrary central character χ_λ with $\lambda \in \Omega$ without change.

5.3 Characteristic variety of a Harish–Chandra bimodule.

By a *Harish–Chandra bimodule*, we mean a $(\mathfrak{g} \times \mathfrak{g}, K)$–module with finite K–multiplicities, where K is the diagonal copy of G in $G \times G$. (Here we use the terminology of [BB3]; so the module is K–equivariant in the strong sense that the differential of the K–action coincides with the action of the diagonal subalgebra \mathfrak{k} of $\mathfrak{g} \times \mathfrak{g}$.) For any finitely generated Harish–Chandra bimodule H with integral central character $\chi_{(\lambda,\mu)}$ given by a pair (λ,μ) of dominant integral weights we define modules L^l resp. L^r of central character χ_λ resp. χ_μ by

$$L^l := M(\mu)^\vee \otimes_{\mathcal{U}(\mathfrak{g}^r)} H \text{ resp. } L^r := M(\lambda)^\vee \otimes_{\mathcal{U}(\mathfrak{g}^l)} H ,$$

where e.g. $\mathfrak{g}^r = 0 \times \mathfrak{g}$ denotes the right copy of \mathfrak{g} in $\mathfrak{g} \times \mathfrak{g}$, and $M(\mu)^\vee$ is the universal (Verma–)module of highest weight μ, considered as a right \mathfrak{g}–module via the principal anti–automorphism of $\mathcal{U}(\mathfrak{g})$. Then $H \longmapsto L^1$ resp. $H \longmapsto L^r$ are *equivalences* of the category of Harish–Chandra bimodules with central character $\chi_{(\lambda,\mu)}$ with the category of all finitely generated (\mathfrak{g},B)–modules with central character χ_λ (resp. χ_μ). For a geometric interpretation (and proof) of this well–known result of Bernstein–Gelfand [BeGe], Joseph [J4], and Enright [E], see [BB3] cf. also 5.9 below. In particular, this establishes bijections between simple objects (up to isomorphism); we shall denote by $H_w^{(\lambda,\mu)}$ for any $w \in W$ the simple Harish–Chandra bimodule corresponding to $L^1 = L\,(w.\lambda)$ resp. to $L^r = L(w^{-1}.\mu)$ (cf. [BB3], 3.4). We denote H^1 the $\mathfrak{g} = \mathfrak{g}^1$–module obtained from H by forgetting the \mathfrak{g}^r–action.

Theorem ([BB3], 5.8, [Gi]): *Let H be any finitely generated Harish–Chandra module of central character $\chi_{(\lambda,\mu)}$, and let L^r be the corresponding (\mathfrak{g},B)–module of central character χ_μ as above. Then*

$$Ch(H^1) = G \times^B V(L^r)\,.$$

Here $V(L^r)$ denotes the associated variety of L^r, that is the support of $gr\,L^r$ in $\mathfrak{g}^* = \mathfrak{g}$ with respect to some good filtration. Of course we obtain an analogous result $Ch(H^r) = G \times^B V(L^1)$ by interchanging left and right.

Corollary: *$Ch(H^1)$ is a union of orbital cone bundles* (terminology 1.7).

In fact, it is enough to show that $V(L^r)$ is a union of orbital cones $C_1,...,C_r$, because then $Ch(H^1)$ is the union of the orbital cone bundles $\mathcal{K}_i = G \times^B C_i$ by the theorem. Let us briefly recall the reason: The localization of L^r on X is a holonomic \mathcal{D}–module and its characteristic variety is a union of closures of conormal bundles of Schubert cells, which project onto orbital cones

under Springer's map $\pi\colon\ T^{*}X \longrightarrow \mathfrak{g}$. On the other hand, π maps $Ch(L^{\mathbf{r}})$ onto $V(L^{\mathbf{r}})$ by [BB3], 1.9.

5.4 Homogeneous Harish–Chandra bimodules.

We denote $d(M) = \dim V(M)$ the Gelfand–Kirillov dimension of a \mathfrak{g}–module M. For a Harish–Chandra bimodule H as in 5.3 it is obvious that $d(H^1) = d(H) = d(H^{\mathbf{r}})$. We call H (left) homogeneous, if $d(M) = d(H)$ for each left submodule $0 \neq M \subset H$. It is easy to see that the following are examples:

a) Each *simple* Harish–Chandra bimodule $H = H_{w}^{(\lambda,\mu)}$ is homogeneous (cf. lemma 5.10).

b) For each *primitive ideal* $J, H = \mathcal{U}(\mathfrak{g})/J$ is a homogeneous

Harish–Chandra bimodule. The $\mathfrak{g} \times \mathfrak{g}$ action is defined by

$$(x,y)u := xu - uy \text{ for } x,y \in \mathfrak{g}, u \in H,$$

see [BB3], §3.

Corollary 1 (notation 5.3): *If* H *is homogeneous, then* $Ch(H^1)$ *is a union of orbital cone bundles* $\mathcal{K}_1,...,\mathcal{K}_r$, *all of the same dimension*

$$\dim Ch(H^1) = \dim X + d(L^{\mathbf{r}}) = \dim X + \tfrac{1}{2} d(H).$$

Proof: By theorem 5.3, it suffices to show that $V(L^{\mathbf{r}})$ is equidimensional of dimension $d(L^{\mathbf{r}}) = \tfrac{1}{2} d(H)$. This follows from the following lemma by a general theorem of Gabber-Kashiwara [Le]. Q.e.d.

Lemma: $L^{\mathbf{r}}$ *is homogeneous of dimension* $d(L^{\mathbf{r}}) = \tfrac{1}{2} d(H)$.

Proof: The equivalence of categories, sending H to L^r, described in 5.3, induces an isomorphism of the lattices of submodules. Let $L' \neq 0$ be a submodule of L^r, and let $H' \subset H$ be the corresponding bisubmodule $H' \neq 0$ in H. Then

$$V(H') = \pi(Ch(H'^1)) = \pi(G \times^B V(L')) = GV(L') = GC_1 \cup ... \cup GC_r \text{ for some}$$

orbital cones $C_1,...,C_r$ by 5.3. But $GC_i = 2 \dim C_i$ for each orbital cone C_i by well–known results of Steinberg and Spaltenstein (see 1.7). Hence $d(H') =$ $\dim V(H'^1) = \dim GV(L') = \max_i \dim GC_i = 2 \max_i \dim C_i = 2 d(L')$. Since H is homogeneous of dimension $d(H)$, it follows now that $d(L') = \frac{1}{2} d(H') = \frac{1}{2} d(H)$.

Q.e.d.

From the above proof, we make the following

Observation: *Each irreducible component K_i of $Ch(H^1)$ maps* (under π) *onto an irreducible component of $V(H^1)$.*

In fact, $\pi(K_i) = \pi(G \times^B C_i) = GC_i$ has dimension $\dim GC_i = 2 \dim C_i$, while $\dim K_i = \dim X + \dim C_i$. Since the last dimension is independent of i by the corollary above, the first one is also, hence $\dim \pi(K_i) = \dim V(H^1) = d(H)$, so $\pi(K_i)$ must be an irreducible component of $V(H^1)$.

Note that in particular this says that $V(H^1)$ is also equidimensional, which is of course a direct consequence of the equidimensionality theory of Gabber-Kashiwara quoted above.

Corollary 2: *If $M \subset M'$ are left submodules $\neq 0$ of a homogeneous Harish-Chandra bimodule H, then*

 a) *$Ch(M)$ is equidimensional of dimension $\dim Ch(H^1)$, and*

 b) *$d(M'/M) < d(H)$ implies $Ch(M) = Ch(M')$.*

Proof:

a) Let H' be the bimodule generated by M. Then H' is finitely generated as a left module, hence is a finite sum of homomorphic images M_j of M. It follows that $Ch(H'^1) = \bigcup_j Ch(M_j) \subset Ch(M)$, hence $Ch(H'^1) = Ch(M)$. Now a) follows from Corollary 1.

b) If M'/M would contribute to the characteristic cycle of M', then $Ch(M'/M)$ would contain an irreducible component of $Ch(M')$, hence $V(M'/M) = \pi(Ch(M'/M))$ would contain an irreducible component of $V(M')$ by the above observation, hence $d(M'/M) = d(M') = d(H)$, contradicting the assumption.

$$Q.e.d.$$

5.5 Characteristic cycle and class of a Harish–Chandra bimodule.

With notations as in 5.3, we have the even stronger result:

Theorem: (cf. [BB3], 5.9): $Ch(H^1) = G \times^B V(L^r)$.

In more detail, this means that if the associated cycle of L^r is

$$V(L^r) = m_1[C_1] + ... + m_r[C_r],$$

with $C_1,...,C_r$ the different irreducible components of $V(L^r)$, then the characteristic cycle of H^1 is

$$Ch(H^1) = m_1[\mathcal{K}_1] + ... + m_r[\mathcal{K}_r]$$

where $\mathcal{K}_i = G \times^B C_i$, and the positive integers m_i are as in (2) (i = 1,...,r).

Corollary 1: *If* H *is homogeneous, then* $P(H^1) = m_1 Q(\mathcal{K}_1) + ... + m_r Q(\mathcal{K}_r)$, *and* $\deg P(H^1) = \dim X - \frac{1}{2} d(H)$.

This follows from 5.1 and Corollary 5.4.1.

Corollary 2: If H *is homogeneous, then* $Ch(H^1)$ *is entirely determined by* $P(H^1)$ (*and conversely*).

In fact, the polynomials $Q(\mathcal{K}_1),...,(\mathcal{K}_r)$ are linearly independent by corollary 1.8.1. Hence $P(H^1)$ determines the multiplicities $m_1,...,m_r$ uniquely.

Corollary 3: If $H \neq 0$, then $P(H^1) \neq 0$.

5.6 Identification with a character polynomial.

Recall that we defined in 3.9 a *character polynomial* q_M for any finitely generated T—equivariant S(u)—module M. Now we consider a finitely generated (\mathfrak{g},B)—module L as in 5.3, and we define its *character polynomial* p_L or $p(L)$ by

$$p_L := q_{grL}' \text{ (or } p_{grL}') \tag{1}$$

where grL denotes the associated graded module with respect to some T—equivariant good filtration of L. It is obvious that such a filtration exists, and that grL is a T—equivariant finitely generated S(u)—module, so q_{grL} is defined. Since the formal characters of L and grL are the same, Corollary 3.9 gives

$$p_L = gr \Delta(u^*) ch(L). \tag{2}$$

We take this formula as an alternative definition of p_L, in terms of the formal character of L, which exhibits the independence of the choice of a filtration.

Theorem (notations 5.3): $P(H^1) = p(L^r)$ (3)

Proof: Let $C_1,...,C_r$ denote the irreducible components of $V(L^r)$ of minimal codimension d, and let $m_1,...,m_r$ denote their multiplicities in $V(L^r)$, so that by (1) and proposition 4.9.2:

$$p(L^r) = q_{grL} = \sum_i m_i p_{C_i}. \tag{4}$$

By theorem 5.4, $K_i := G \times^B C_i$, for $i = 1,...,r$, are the irreducible components of $Ch(H^1)$ of minimal codimension d in T^*X, so

$$P(H^1) = \sum_i m_i \, Q(K_i) \qquad (5)$$

by 5.1 and theorem 5.5. Now we have by 4.6, resp. theorem 4.7.1, resp. 4.5

$$Q(K_i) = Q_G(K_i) = Q_G(K_i) = P_{C_i} = q_{C_i}$$

for each $i = 1,...,r$. So the sums in (4) and (5) are equal term by term, and (3) follows. Q.e.d.

Remark: This proof of the theorem is based on the harmonicity of the character polynomial q_{C_i} (theorem 4.7.1). We shall give in 5.9 an *alternative proof,* based more directly on the harmonicity of the character polynomial $p(L^r)$, which seems more satisfactory.

Corollary: *For a simple Harish–Chandra bimodule, we have* (notation (5.3)

$$P(H_w^{(\lambda,\mu),l}) = p(L(w^{-1} \cdot \mu)),$$
$$P(H_w^{(\lambda,\mu),r}) = p(L(w \cdot \lambda)) \ .$$

5.7 Harmonicity of character polynomials.

Let L be any finitely generated (\mathfrak{g},B)–module of central character χ_μ. Let us recall from the theory of highest weight modules [Di], [Ja] that within an appropriate Grothendieck group, L is expressible as an integer linear combination of Verma modules $M(\nu)$ of highest weight $\nu = w.\mu$, $w \in W$, say

$$[L] = \sum_{w \in W} a_w(L) \, [M(w.\mu)], \qquad (1)$$

where the integer coefficients $a_w(L)$ are uniquely determined. We list some properties of the character polynomial $p(L)$, which are well–known, but briefly reproved here for convenience of the reader.

Proposition:

a) $p(L)$ *is homogeneous of degree* $a := \dim u - d(L)$.

b) $p(L) = \frac{1}{a!} \sum_{w \in W} a_w(L) \, (w.\mu)^a$ (2)

c) a *is also characterized as the smallest integer* ≥ 0 *that makes the right hand side of* (2) *nonzero.*

d) *Let* $a(L)$ *denote the group ring element* $\sum_{w \in W} a_w(L) w$. *Then* $a(L) \frac{1}{j!} (\mu+\rho)^j$ *is zero for* $j < a$, *and is* $p(L)$ *for* $j = a$.

e) $p(L)$ *is harmonic.*

Proof:

a) By theorem 3.10, the degree of the character polynomial equals to the codimension of the support of grL in u^*, hence a) follows from $d(L) = \dim V(L) = d(grL)$.

b),c) Since the formal character of a Verma module is obviously given by $ch(M(\nu)) = \Delta(u^*)^{-1} e^{\nu}$ [Di], equation (1) gives

$$\Delta(u^*) ch(L) = \sum_{w \in W} a_w(L) e^{w.\mu}. \qquad (3)$$

Writing out the exponentials as power series, we obtain

$$[\Delta(u^*) ch(L)]^j = \sum_{w \in W} a_w(L) \frac{1}{j!} (w.\mu)^j$$

for the homogeneous term of degree j of (3), for all $j \geq 0$. Now b) and c) follow from a).

d) follows from a), b), c) by binomial development of $(w.\mu)^j = (w(\mu+\rho)-\rho)^j$ using induction on j.

e) Let Q be a constant coefficient differential operator on t which is W invariant, say Q homogeneous of degree $d > 0$. We have to prove $Qp(L) = 0$. This is clear for $\deg Q > \deg p(L)$. So let $d \leq a$.

By Leibniz's rule, we have $Q\lambda^a = c.Q(\lambda)\lambda^{a-d}$ for some scalar c independent on $\lambda \in t^*$, where we consider Q as a polynomial function on t^*. Using the formula for $p(L)$ given in d), we obtain

$$Qp(L) = Q \mathfrak{a} (L) \frac{1}{a!} (\mu+\rho)^a = \mathfrak{a} (L) Q \frac{1}{a!} (\mu+\rho)^a$$

$$= \frac{c}{a!} Q(\mu+\rho) \mathfrak{a} (L) (\mu+\rho)^{a-d} = 0 ,$$

where the second equation comes from W invariance of Q, and the last comes from d), since $a-d$ is strictly smaller than a. Q.e.d.

5.8 Equivariant characteristic class for a Harish–Chandra bimodule.

Let H be a finitely generated Harish–Chandra bimodule with central character $\chi_{(\lambda,\mu)}$ as in 5.3. Then H has a localization \mathcal{X} on the flag variety $Z = X \times X$ of the group $G \times G$. By definition (cf. [BeBe], [BB3]), \mathcal{X} is the coherent $(\mathcal{D}_Z^{(\lambda,\mu)}, K)$–module

$$\mathcal{X} := \mathcal{D}_Z^{(\lambda,\mu)} \otimes_{\mathcal{U}(\mathfrak{g}\times\mathfrak{g})} H,$$

where $\mathcal{D}_Z^{(\lambda,\mu)}$ denotes the sheaf of rings of differential operators with coefficients in the line bundle $\mathcal{L}(\lambda,\mu)$ on Z given by (λ,μ). Now let $p^1: = X \times X \longrightarrow X$ denote projection onto the left copy of X. Then the direct image sheaf $p^1_* \mathcal{X}$ is a coherent \mathcal{D}_X^λ–module, which still carries an action of $G \simeq K$. Note, however, that while \mathcal{X} is K–equivariant in the strong sense of [BB3], 2.2, the module $p^1_*\mathcal{X}$ is only "weakly G–equivariant" in the sense of 2.12. The following is easy to see (cf. [BB3], 5.10a) resp. 5.5a)):

Lemma:

a) *\mathcal{X} admits a K–equivariant good filtration $(\mathcal{X}_n)_{n\in\mathbb{Z}}$.*

b) *This induces a G–equivariant good filtration $(p^1_*\mathcal{X}_n)_{n\in\mathbb{Z}}$ on $p^1_*\mathcal{X}$.*

With respect to such a filtration, the associated graded module $\operatorname{gr} p_*^1 \mathcal{I}$ is a G–equivariant \mathcal{O}_{T^*X}–module, and hence defines a class in $K_G(T^*X)$, which is independent of the choice of filtration by Lemma 2.12. Under the composition of the maps $K_G(T^*X) \longrightarrow K_G(X) \longrightarrow R(T) \subset \hat{S}(t^*)$, this class determines a power series on t, of lowest degree homogeneous term $Q_G(\operatorname{gr} p_*^1 \mathcal{I})$ (see definition 4.5). Now we define

$$P_G(H) := Q_G(\operatorname{gr} p_*^1 \mathcal{I}).$$

Proposition: $P(H^1)$ *is the harmonic part of* $P_G(H)$.

Proof: One has to observe that $p_*^1 \mathcal{I}$ is isomorphic to the localization of H^1 on X (cf. [BB3], proposition 5.4). Then

$$P(H^1) = Q(\operatorname{gr} p_*^1 \mathcal{I}) = Q_G(\operatorname{gr} p_*^1 \mathcal{I}) = P_G(H)$$

follows from proposition 4.6. Q.e.d.

5.9 Alternative proof of identification with character polynomials (5.6).

With notations as in 5.8, let us compute $P_G(H)$, using the functor $H \longmapsto L^r$ described in 5.3. Let us first recall the *geometric interpretation of this functor* (cf. [BB3], 3.6): The inclusion of X into $Z = X \times X$ as the right copy $X^r = \{B\} \times X$ is denoted i_r. The restriction $\mathcal{L}^r := i_r^* \mathcal{I}$ of the $\mathcal{D}_Z^{(\lambda,\mu)}$–module \mathcal{I} to $X^r = X$ is a B–equivariant \mathcal{D}_X^μ–module, which is canonically isomorphic to the localization of L^r on X, and the functor $i_r^* : \mathcal{I} \longmapsto \mathcal{L}^r$ establishes an equivalence of the category of coherent $(\mathcal{D}_Z^{(\lambda,\mu)}, K)$–modules with the category of coherent (\mathcal{D}_X^μ, B)–modules. We note that $L^r \cong \Gamma(X, \mathcal{L}^r)$ (cf. loc.cit.).

Lemma (cf. [BB3], 5.10): *A* K*–equivariant good filtration on* \mathcal{H} *induces a* B*–equivariant good filtration on* $i_r^* \mathcal{H} = \mathcal{L}^r$. *With respect to such filtrations, we have a* B*–equivariant isomorphism of graded* B*–equivariant* \mathcal{O}_u*–modules*

$$j^*(\mathrm{gr}\ p_*^1\ \mathcal{H}) \simeq \mathrm{gr}\ \Gamma(X, i_r^* \mathcal{H}) \simeq \mathrm{gr}\ L^r .$$

Here $j : u \hookrightarrow T^*X$ denotes the inclusion of the cotangent space at the base point; also we notationally do not distinguish between the $\mathcal{O}(u)$–module $\mathrm{gr}\ L^r$ and the corresponding sheaf of \mathcal{O}_u–modules, since u is affine.

Theorem (notation 5.3): $P_G(H) = p(L^r)$.

Proof: By definition, $P_G(H) = Q_G(\mathrm{gr}\ p_*^1 \mathcal{H})$, with notations as in 5.9. By proposition 4.5, $Q_G(\mathrm{gr}\ p_*^1 \mathcal{H})$ is the character polynomial of $j^*(\mathrm{gr}\ p_*^1 \mathcal{H})$. By the lemma, this coincides with the character polynomial of $\mathrm{gr}\ L^r$, hence with $p(L^r)$ by definition 5.6(1). $\hspace{2cm}$ Q.e.d.

As a *corollary*, we obtain the following *alternative proof of theorem* 5.6:

$$P(H^1) = P_G(H) = p(L^r) = p(L^r)$$

by proposition 5.8, the theorem above, and proposition 5.7 e).

$\hspace{11cm}$ Q.e.d.

Remark: In all preceding considerations, we may obviously interchange left and right sides to obtain analogous results. For instance, we define the "right" equivariant characteristic class of H by $P_G(H^r) := Q_G(\mathrm{gr}\ p_*^r \mathcal{H})$, where $p^r : X \times X \longrightarrow X$ denotes projection onto the right copy, in complete analogy to the "left" one, $P_G(H^1) = P_G(H)$ in 5.8, and then we obtain

$$P_G(H^r) = p(L^1) = P(H^r)$$

in analogy to the theorem and corollary above. In the sequel, we leave the

statements of "right analogues" to the reader, and consider only *left* characteristic cycles and classes $P(H^1)$; we shall eventually even give up to write the "1", in order to avoid too clumsy notation in our formulas.

5.10 Some non–commutative algebra.

The *Goldie rank* of a module M, denoted rk M, is the maximal number r such that M contains a direct sum of r non–zero submodules. A non–zero module is called *uniform*, if any two non–zero submodules have non–zero intersection, i.e., if rk M = 1. If M is noetherian, then clearly rk M < ∞, and M contains a direct sum of (rk M) uniform submodules; moreover, it is a fact from general ring theory that any direct sum within M of uniform submodules (necessarily of \leq (rk M) terms) can be extended to a direct sum within M of rk M uniform terms (see e.g. [Go], Theorem 1.07).

Proposition: *Let* H *be a simple Harish–Chandra bimodule. Let* U *be a uniform submodule of* H^1. *Then*

a) H^1 *contains a direct sum* M *of* rk H^1 *copies of* U.

b) $d(H^1/M) < d(H)$.

Corollary: a) $Ch\,(H^1) = $ rk $H^1\ Ch(U)$

b) $Ch\,(U)$ *is independent of the choice of* U.

Note that corollary a) is immediate from the proposition by corollary 5.4.2, and b) follows from a) on dividing by the positive integer rk H^1.

Lemma:

a) H^1 *is homogeneous of* \mathcal{GK}*–dimension* $d(H) = d(A)$ *where* $A = U(\mathfrak{g})/\mathrm{Ann}\ H^1$.

b) *Every proper homomorphic image of* U *has smaller* \mathcal{GK}*–dimension.*

Proof of the lemma:

a) If $N \subset H^1$ has $d(N) < d(H)$, then $d(NU(\mathfrak{g})) \leq d(N) < d(H)$, so $NU(\mathfrak{g})$ would be a proper bisubmodule of H, hence $N = 0$ by simplicity of H. The inequality $d(H) \leq d(A)$ is trivial, and the opposite inequality follows e.g. from the fact that A^1 embeds into a finite direct sum of copies of H^1 (cf. e.g. [BB3], 4.10).

b) Let $\overline{} : U \longrightarrow \overline{U}$ be a homomorphism with kernel $N \neq 0$. Let $u \in U$, and $L = \{a \in A \mid a\overline{u} = 0\}$. We claim that this is an essential left ideal of A, i.e., that every left ideal $L' \neq 0$ meets L non–trivially. In fact, suppose $L' \cap L = 0$. Then $L' \to L'\overline{u}$ is injective, so $L' \longrightarrow L'u$ is injective, and $L' \cap N = 0$, contradicting uniformicity of U. Now it follows from [BGR], 2.7 that L contains a nonzerodivisor s of A, since A is a prime noetherian ring. Now we conclude that $d(A\overline{u}) = d(A/L) \leq d(A/As) \leq d(A) - 1$ by the argument given in [BK2], 3.4 or by [B3], 1.3. By a) this proves $d(\overline{U}) < d(A) = d(U)$.

Q.e.d.

Proof of the proposition: Here we write H as a left–right $U(\mathfrak{g})$–bimodule. Let $0 \neq E \subset H$ be a finite dimensional K–submodule; then $H = U(\mathfrak{g})EU(\mathfrak{g}) = U(\mathfrak{g})E$, so H^1 is finitely generated (hence noetherian). By simplicity of H, $UU(\mathfrak{g}) = H$, hence $UF = H$ for some finite dimensional subspace $F \subset U(\mathfrak{g})$ by noetherianness of H^1. For each $f \in F$, Uf is a homomorphic image of U. If $Uf \neq 0$, then $d(Uf) = d(H) = d(U)$ by homogeneity (lemma a)), so $Uf \cong U$ by lemma b). We have proved that $H = UF$ is a finite sum of copies of U. Let $M = Uf_1 + ... + Uf_r$ be a maximal sub–sum $(f_1,...,f_r \in F)$ such that the sum is direct. It remains to

prove that $d(H/M) < d(H)$, and that $r = \text{rk } H^1$. Suppose $d(H/M) = d(H)$. Then since H^1 is a finite sum of submodules of the type $U.f$ $(f \in F)$, $d(H/M)$ is the maximum of $d(\overline{Uf})$, where $\overline{}$ denotes the quotient map $H \longrightarrow H/M$. Hence $d(H/M) = d(H)$ implies $d(\overline{Uf}) = d(H)$ for some $0 \neq f \in F$, so Uf maps isomorphically onto \overline{Uf} by lemma b), so $Uf \cap M = 0$, contradicting the maximality of M. Hence $d(H/M) < d(H)$ is proved. Clearly $r \leq \text{rk } H^1$. If $r <$ $\text{rk } H^1$, then by the remark preceding the proposition, there exists a submodule $0 \neq U' \subset H^1$ such that $U' \cap M = 0$. Since this implies $d(H/M) \geq d(U') = d(H)$ by homogeneity of H^1, this contradicts $d(H/M) < d(H)$. Hence we must have $r = \text{rk } H^1$.

<div align="right">Q.e.d.</div>

5.11 Definition of the polynomials P_w.

Lemma: *Let* U *be a uniform left submodule of* $H_w^{(0,0)}$ *for some* $w \in W$. *Then for every pair* $\lambda, \mu \in \Omega$, *the module* $T_0^\lambda U$ *(notation 5.2) is isomorphic to a uniform submodule of* $H_w^{(\lambda,\mu)}$.

Proof: We have $T_0^\lambda H_w^{(0,0),1} = (T_{(0,0)}^{(\lambda,0)} H_w^{(0,0)})^1 = H_w^{(\lambda,0),1}$, so $U' := T_0^\lambda U$ is a left submodule of $H_w^{(\lambda,0)}$; since T_0^λ is an equivalence of categories, it preserves the lattice of submodules up to isomorphism, so U uniform implies U' uniform.

Now we have $H_w^{(\lambda,\mu)} = T_{(\lambda,0)}^{(\lambda,\mu)} H_w^{(\lambda,0)}$ is a direct summand of $H_w^{(\lambda,0)} \otimes E(0,\mu)$ (notation 5.2); let $p: H_w^{(\lambda,0)} \otimes E(0,\mu) \longrightarrow H_w^{(\lambda,\mu)}$ be the projection map. As a left module, $H_w^{(\lambda,0)} \otimes E(0,\mu)$ is a finite direct sum of copies of $H_w^{(\lambda,0)}$, hence a finite sum of copies of U' (by the first part of the argument in 5.10). So $p(U'') \neq 0$ for at least one of these copies, U'' say. But then $d(p(U'')) = d(H_w^{(\lambda,\mu)}) =$

$d(H_w^{(\lambda,0)}) = d(U')$ by homogeneity (lemma 5.10a)), hence $p(U'') \simeq U'' \simeq U'$ by lemma 5.10 b). Q.e.d.

Theorem: *Let $w \in W$. Take any two pairs (λ,μ), (λ',μ') of dominant integral weights. Let U resp. U' be uniform left submodules of $H_w^{(\lambda,\mu)}$ resp. $H_w^{(\lambda',\mu')}$. Then $Ch(U) = Ch(U')$, and $P(U) = P(U')$.*

Proof: The characteristic cycle $Ch(U)$ is independent of the choice of U by corollary 5.10.b). Making a choice as in the lemma, we see that this cycle is independent of μ, and also of λ by 5.2. The second claim follows by 5.1

 Q.e.d.

Definition: *For each $w \in W$, we denote $P_w := P(U)$ the polynomial uniquely determined by the theorem.*

Corollary: $P(H_w^{(\lambda,\mu)}) = rk\ H_w^{(\lambda,\mu),1}\ P_w$.

This follows from corollary 5.10 a) by 5.1. In particular, we note that P_w is a homogeneous harmonic polynomial of degree $a = a(w)$.

5.12 Relation to primitive ideals.

Let J be a primitive ideal of $U(\mathfrak{g})$, and $A := U(\mathfrak{g})/J$. We note that the left resp. right modules have the same Goldie rank, denoted $rk\ A$, also called the Goldie rank of J. This number has the following well–known alternative interpretation: Let S denote the set of non–zero divisors of A; by Goldie's theorem, A admits a ring of fractions $S^{-1}A$, which is simple artinian, and hence

isomorphic to the ring of n by n matrices over some skew field (Wedderburn–Artin theorem); then $n = rk\ A$.

Now assume J of integral central character χ_λ. We recall that

$$J = I(w.\lambda) := Ann\ L(w.\lambda) \tag{1}$$

for some $w \in W$ (Duflo's theorem [D]), and that a simple Harish–Chandra module $H_w^{(\lambda,\mu)}$ has left resp. right annihilator

$$Ann\ H_w^{(\lambda,\mu),1} = I(w.\lambda) \quad resp. \quad Ann\ H_w^{(\lambda,\mu),r} = I(w^{-1}.\mu). \tag{2}$$

(Note that we are still considering left–right $U(\mathfrak{g})$–modules.) The last result (due to Joseph [J5]) is easily deduced from the equivalences stated in 5.3 (cf. also [Ja], 7.9).

Theorem: *Let* $A := U(\mathfrak{g})/I(w.\lambda)$, *where* $w \in W$, $\lambda \in \Omega$. *Let* U *be a uniform left ideal of* A. *Then*

 a) $P(A) = rkA\ P(U)$

 b) $P(U) = P_w$ (definition 5.11).

Corollary 1: $P_w = P_y$ *for* $w,y \in W$ *in the same left cell.*

Here two Weyl group elements w,y are said to be in the same *left cell,*if $I(w.\lambda) = I(y.\lambda)$ for one (and hence for all) $\lambda \in \Omega$. Note that the corollary follows from the theorem because $P(A)/rkA$ depends only on the ideal $I(w.\lambda)$.

Notation: The corollary justifies the notation $P_J = P_w$ if w is any element in W such that $J = I(w.\lambda)$.

Proof:

a) Considered as a Harish–Chandra bimodule, A has finite length, and hence a socle $H \neq 0$, which obviously must be simple by primality of A (as a ring), so

$$\text{socle}(A) = H \cong H_v^{(\lambda,\lambda)} \tag{3}$$

for some $v \in W$. Also $d(A/H) < d(A)$ by [BK2], Satz 3.4, and hence $P(A) = P(H)$ by corollary 5.4.2 and 5.1. Now let $U' := U \cap H$; then also $d(U/U') \leq d(A/H) < d(A)$, so that we again conclude $P(U) = P(U')$ by loc.cit. We also must have $\text{rk } A = \text{rk } H^1$, because otherwise $\text{rk } H^1 < \text{rk } A$, and then we could find a uniform submodule of A^1 injecting into A/H by the remark preceding proposition 5.10, which would contradict $d(A/H) < d(A)$ (using homogeneity of A^1). Applying corollary 5.10a) (plus 5.1) to H and U' (which is of course uniform), we may now conclude

$$P(A) = P(H) = \text{rk } H^1 \, P(U') = \text{rk } A \, P(U).$$

b) Note that (3) gives us $P(U) = P(U') = P_v$, but only for a particular element v in the left cell of w, so this is not enough to prove b). Instead, we proceed as follows. We embed A^1 into a finite direct sum of copies of $H_w^{(\lambda,0)}$ (using (2) and [BB3], lemma 4.10). Projecting onto a suitable one of these copies, we get a homomorphism of U' into $H_w^{(\lambda,0),1}$ with nonzero image. Since $H_w^{(\lambda,0),1}$ is homogeneous of dimension $d(A)$ (use (2) and lemma 5.10a)) we have $d(pU') = d(A)$, so $pU' \cong U'$ by lemma 5.10b). We have proved that $H_w^{(\lambda,0),1}$ contains a copy of U', so (by definition 5.11)

$$P(U) = P(U') = P_w. \qquad \text{Q.e.d.}$$

Remark 1: We have also seen in the proof that:

Corollary 2: *There exists* $v \in W$ *such that* $I(w.\lambda) = I(v.\lambda)$,

$$P(U(\mathfrak{g})/I(w.\lambda)) = P(H_v^{(\lambda,\lambda)}), \qquad (4)$$

and

$$\mathrm{rk}\ U(\mathfrak{g})/I(w.\lambda) = \mathrm{rk}\ H_v^{(\lambda,\lambda),1} . \qquad (5)$$

Remark 2: The choice of v above in the left cell of w is uniquely determined by (3). We further *note that* $v^2 = 1$ (Duflo [D], Proposition 9). Since $H_w^{(\lambda,\lambda)} = T_{(0,0)}^{(\lambda,\lambda)} H_v^{(0,0)}$ is the simple socle of $U(\mathfrak{g})/I(w.\lambda) = T_{(0,0)}^{(\lambda,\lambda)} (U(\mathfrak{g})/I(w.0))$, it can be concluded that v is uniquely determined by w, i.e. independent of λ (cf. [Ja], 7.11). Logically, we shall not have to use this fact here, because we anyway know by corollary 1 that P_v is independent of λ.

Remark 3: We note that in general (4) does not hold with w in place of v. Although we do have (even for arbitrary $\mu \in \Omega$)

$$Ch(U(\mathfrak{g})/I(w.\lambda)) = Ch\ (H_w^{(\lambda,\mu)}) \ (= G \times^B V\ (L(w^{-1}.\mu))),$$

the corresponding equation for characteristic cycles (or equality for characteristic classes, Cor. 5.5.2) is only true up to a proportionality factor, as is seen by combining theorems 5.11 and 5.12. In fact, the factor is the ratio of two Goldie–ranks and is a function of λ,μ, and w in the left cell. We shall analyze the "Goldie rank functions" in the last sections of this chapter (5.15–5.18). But before doing this, let us first make some remarkable applications of theorems 5.11 resp. 5.12.

5.13 Irreducibility of Joseph's Weyl group representations.

Theorem 1: *Each polynomial* P_w *(w ∈ W)* *generates an irreducible* W *submodule of* S(t).

We shall see (in 5.17) that our polynomial P_w is proportional to \tilde{p}_w in Joseph's notation [J3], [J1], so this theorem 1 and also theorem 2 below are just reformulations of theorems of Joseph [J3]. Our point to make here is that theorem 1 is an easy consequence of the result (5.11) underlying our definition of P_w. The argument is the same as in [Ja], 14.10, but we make it explicit here for convenience of the reader:

Proof: We consider the group ring element $a_w := a\,(L(w^{-1}.\mu)) \in k[W]$, and the degree $a = a(w)$ as defined in 5.7. Then we have

$$a_w \frac{1}{a!}\,\mu^a = p(L(w^{-1}.(\mu-\rho))) = P(H_w^{(\lambda,\mu-\rho)}) \in \mathbb{N}P_w \subset kP_w$$

by 5.7d), theorem 5.6, and corollary 5.11, for all $\mu \in \Omega + \rho$. Using that $\Omega + \rho$ is Zariski dense in $t^* = t$, and that the powers μ^a ($\mu \in t$) span $S^a(t)$, we conclude that the linear operator a_w projects all of $S^a(t)$ onto the line kP_w. (In particular, P_w is an eigenvector.) On the other hand, $S^a(t)$ splits into a direct sum of irreducible W submodules, each of which is projected into itself by a_w. So a_w must kill all but one irreducible summand, E say. Then $P_w \in a_w E \subset E$, so P_w generates E. Q.e.d.

Remark: The argument shows simultaneously that E has multiplicity one in $S^a(t)$ resp. zero in $S^j(t)$ for $j < a$.

Although we do not use it in the sequel, let us restate here — for the sake of completeness — also Joseph's result about the *classification of primitive ideals* [J3].

Theorem: *If* $J_1,...,J_r$ *are all primitive ideals of a given central character* χ_λ *$(\lambda \in \Omega)$ corresponding to a given irreducible* W *module* E *by theorem 1, then* $P_{J_1},...,P_{J_r}$ *(notation 5.12) form a basis of* E.

This is now derived from Vogan's characterization of the order relation of primitive ideals [V1], as exposed e.g. in chapters 7 and 14 of Jantzen's book [Ja].

5.14 Irreducibility of associated varieties of primitive ideals.

Theorem 1 [BB1]: *Each primitive ideal* J *with integral central character has an irreducible associated variety* $V(U(\mathfrak{g})/J)$.

Since this variety is obviously G–invariant, and contained in the cone \mathcal{N} of nilpotent elements, it is then necessarily the closure of a single nilpotent orbit, say \mathcal{O}_x.

Proof [BBM2], [Gi]: Let $J = I(w.\lambda)$ with $w \in W$, $\lambda \in \Omega$. Then the (left) characteristic class $P(U(\mathfrak{g})/J)$ is proportional to P_w by theorem 5.12, hence generates an irreducible W–module by theorem 5.14. On the other hand, we have

$$P(U(\mathfrak{g})/J) = m_1 \, Q(\mathcal{K}_1) + ... + m_r \, Q(\mathcal{K}_r) , \qquad (*)$$

as in 5.5.1, where \mathcal{K}_i is an orbital cone bundle of codimension $a = a(w) \ (=\deg P_w)$ in T^*X, and $0 \neq m_i \in \mathbb{N}$ is also the multiplicity of \mathcal{K}_i in the characteristic cycle $Ch(U(\mathfrak{g})/J)$ (by 5.5.2), for each $i = 1,...,r$. By theorem 1.8, each $Q(\mathcal{K}_i)$ generates an irreducible W submodule of $S^a(t)$ equivalent to Springer's representation ρ_{u_i}, if \mathcal{O}_{u_i} is the nilpotent orbit determined by \mathcal{K}_i (notation 1.8), i.e. if $\pi(\mathcal{K}_i) = \overline{\mathcal{O}}_{u_i}$. Suppose $\mathcal{O}_{u_i} \neq \mathcal{O}_{u_1}$ for some i. Then the corresponding Springer representations

ρ_{u_i}, ρ_{u_1} would be inequivalent by Springer's theory (cf. [BM1]), so they both occur in the cyclic W–module generated by (*), contradicting the irreducibility of this module (5.16). Hence the nilpotent orbits \mathcal{O}_{u_i} are all the same \mathcal{O}_x. Now by [BB3], 1.9,

$$V(U(\mathfrak{g})/J) = \pi \, Ch(U(\mathfrak{g})J)^1 = \pi(\underset{i}{U} \mathcal{K}_i) = \overline{\mathcal{O}}_x .$$

<div align="right">Q.e.d.</div>

Note that we have proved simultaneously:

Theorem 2 (Barbasch–Vogan)[BV1],[BV2]): *Joseph's Weyl group representation generated by* $P(U(\mathfrak{g})/J$ *is equivalent to Springer's Weyl group representation* ρ_x *corresponding to the dense nilpotent orbit in* $V(U(\mathfrak{g})/J)$.

Remark: Theorems 1 resp. 2 had first been verified by the first two authors [BB1] resp. Barbasch–Vogan [BV1], [BV2] using case by case arguments; conceptual proofs were then given by Joseph [J2], Kashiwara–Tanisaki [KT], the first two authors [BB3], resp. by Hotta–Kashiwara [HK]. The argument given in the present subsection appeared independently in Ginsburg's [Gi] and in [BBM2].

5.15 Evaluation of character polynomials.

In the notation of 5.7, we define a W × W matrix of integers by

$$a(w,y) := a_y(L(w.\lambda)) \qquad \text{for all } w,y \in W.$$

By a well–known "translation principle" (due to Jantzen), these integers are independent of $\lambda \in \Omega$, as is the degree

$$a := a(w) := \dim \mathfrak{u} - d(L(w.\lambda))$$

in 5.7.

Remark: Let us mention that in the notation of Kazhdan–Lusztig [KL],

$$a(w.y) = (-1)^{l(w)-l(y)} P_{y,w}(1), \tag{1}$$

where the Kazhdan–Lusztig polynomials $P_{y,w}$ are defined by a purely combinatorial recursion formula. We recall that (1) was conjectured by Kazhdan–Lusztig, and proved by Beilinson–Bernstein [BeBe] and Brylinski-Kashiwara [BKa]. As a consequence, these numbers can be effectively calculated on a computer.

Definition: For each $w \in W$, we define polynomial functions p_w resp. \tilde{p}_w on $t \times t^*$ resp. t by

$$p_w(\xi,\eta) = \sum_{y \in W} a(w,y) <y(\xi+\rho),\, \eta+\rho>^a \tag{2}$$

resp.

$$\tilde{p}_w(\xi) = p_w(\xi,0); \tag{3}$$

in this context, we shall identify t with t^* by the usual scalar product. Then our p_w, \tilde{p}_w coincide with those considered in [J1], 5.1, up to a shift by ρ.

Lemma: *For* $\xi,\eta \in t = t^*$, *and each* $w \in W$ *we have*

$$p_w(\xi,\eta) = p_{w^{-1}}(\eta,\xi).$$

Proof: This follows from (2) by W–invariance and symmetry of the scalar product, using the following property of the coefficients $a(w,y)$:

$$a(w,y) = a(w^{-1},y^{-1}) \qquad \text{for all } w,y \in W.$$

This property follows e.g. by composing the equivalence functors $L^1 \longmapsto H \longmapsto L^r$ (notation 5.3), sending $L(w.\lambda)$ to $L(w^{-1}.\lambda)$, and $M(w.\lambda)$ to $M(w^{-1}.\lambda)$, hence identifying $a(w,y)$ with $a(w^{-1},y^{-1})$. Q.e.d.

Proposition: *For all* $w \in W$, $\lambda, \mu \in \Omega$, $\nu \in t^* = t$, *we have*

$$P(H_w^{(\lambda,\mu)})(\nu+\rho) = p(L(w^{-1}.\mu))\,(\nu+\rho) = \tfrac{1}{a!}\, p_w(\nu,\mu).$$

Proof: The first equation is corollary 5.6, the second is proposition 5.7d), in combination with the lemma. \qquad Q.e.d.

Corollary: $p_w(\lambda,\mu)$ *is a positive integer for all* $\lambda, \mu \in \Omega$.

Proof: They are integers by definition. They are positive, because the character polynomial $p(L(w^{-1}.\mu))$ takes only strictly positive values on the set $\Omega + \rho$ of regular dominant integral weights, by 3.11. \qquad Q.e.d.

5.16 Computation of Goldie ranks

From corollary 5.11, we obtain that for all $\lambda, \mu \in \Omega$

$$\frac{P(H_w^{(\lambda,\mu)})}{\mathrm{rk}\ H_w^{(\lambda,\mu)}} = P_w = \frac{P(H_w^{(0,0)})}{\mathrm{rk}\ H_w^{(0,0)}}$$

(From now on, let rk denote always *left* Goldie rank, to slightly simplify notation.) By proposition 5.15, we get from this equality of polynomials an equality of values, so

$$p_w(\nu,\mu)/\mathrm{rk}\ H_w^{(\lambda,\mu)} = p_w(\nu,0)/\mathrm{rk}\ H_w^{(0,0)} \tag{1}$$

for all $\nu \in t^*$. Using corollary 5.15, we get at least for all $\nu \in \Omega$

$$\mathrm{rk}\ H_w^{(\lambda,\mu)} = \mathrm{rk}\ H_w^{(0,0)}\, \frac{p_w(\nu,\mu)}{p_w(\nu,0)}\ . \tag{2}$$

Taking $\nu = 0$, this gives:

Theorem: *The left Goldie rank of a simple Harish–Chandra bimodule is given by*

$$\mathrm{rk}\ H_w^{(\lambda,\mu)} = c_w \cdot \tilde{p}_{w^{-1}}(\mu) \tag{3}$$

for all $\lambda, \mu \in \Omega$, $w \in W$, *where* c_w *is a positive rational constant.*

In fact, $c_w = rk\ H_w^{(0,0)}/p_w(o,o)$. Note that the polynomial thus describing the left Goldie ranks by (3) for given w is uniquely determined by (3), since Ω is Zariski dense in t^*.

Remark. It follows now from corollary 5.12.2, that also the Goldie ranks of primitive ideals are given by polynomial functions. More precisely,

$$rk\ U(\mathfrak{g})/I(w.\lambda) = c_v \cdot \tilde{p}_{v^{-1}}(\mu)\ ,$$

where $v(= v^{-1})$ is chosen as in 5.12.2. This is a famous result of Joseph [J3], of which we shall obtain a more complete version in 5.18 below.

5.17 Joseph–King factorization of polynomials p_w.

From 5.16 (1), we get more generally

$$rk\ H_w^{(\lambda,\mu)} = c_w(\nu_0) \cdot p_w(\nu_0,\mu)\ ,\tag{1}$$

for any $\nu_0 \in \Omega$, where $c_w(\nu_0)$ is a positive constant $(= rk\ H_w^{(0,0)}/p_w(\nu_0,o))$, independent of λ,μ. Now we use 5.16 (1) for a second choice of μ, say $\mu' \in \Omega$, to get

$$p_w(\nu,\mu)\ /\ rk\ H_w^{(\lambda,\mu)} = p_w(\nu,\mu')\ /\ rk\ H_w^{(\lambda,\mu')}\tag{2}$$

for all $\nu,\mu,\mu' \in \Omega$. Going with (1) into (2) twice, and cancelling the nonzero factor $c_w(\nu_0)$ on both sides, we obtain

$$p_w(\nu,\mu)\ /\ p_w(\nu_0,\mu) = p_w(\nu,\mu')\ /\ p_w(\nu_0,\mu')$$

for all $\nu,\nu_0,\mu,\mu' \in \Omega$. Using Zariski density of Ω in t^*, we derive the following remarkable polynomial identity:

Proposition: $p_w(\xi,\eta)\ p_w(\xi',\eta') = p_w(\xi,\eta')\ p_w(\xi',\eta)\ .$

Taking the special case $\xi' = \eta' = 0$, we obtain the following factorization

Corollary 1 (Joseph–King, cf. [J1], 5.1 [Ki]):

$$P_w(\lambda,\mu) = \frac{P_w(\lambda,0)\cdot P_w(0,\mu)}{P_w(0,0)} = c \cdot \tilde{p}_w(\lambda)\,\tilde{p}_{w^{-1}}(\mu).$$

This factorization, up to the constant factor $c = p_w(0,0)^{-1}$, was given in [J1], 5.1; a direct proof by combinatorics of the Kazhdan–Lusztig polynomials does not seem to be known. Now we may rewrite proposition 5.15 as follows:

Corollary 2: $P(H_w^{(\lambda,\mu)})(\nu+\rho) = p(L(w^{-1}.\mu))(\nu+\rho) = \frac{c}{a!}\,P_{w^{-1}}(\mu)\,\tilde{p}_w(\nu).$

Corollary 3: *The polynomial* P_w *(5.11) coincides with* p_w, *up to a positive integer constant factor, and a shift by* ρ; *more precisely:*

$$\tilde{p}_w(\nu) = \mathrm{rk}\,H_w^{(0,0)}\,a!\,\tilde{p}_w(0)\,P_w(\nu+\rho)\,.$$

This formula follows from corollary 2, using corollary 5.11, and theorem 5.16.

Remark: At this point, it becomes clear that our formulation of Joseph's irreducibility theorem (5.13) is equivalent to Joseph's.

Corollary 4:
$$P_w(\rho) = \frac{1}{a!}\cdot\frac{1}{\mathrm{rk}\,H_w^{(0,0)}}$$

Corollary 5 (Joseph): *For* $w,y \in W$ *in the same left cell,* \tilde{p}_w *is proportional to* \tilde{p}_y.

In fact, $P_w = P_y$ by corollary 5.11.1, and of course $a(w) = a(y)$, so

$$\tilde{p}_w = \frac{\tilde{p}_w(0)}{\tilde{p}_y(0)}\cdot\tilde{p}_y$$

by corollary 3.

Remark: The converse of this statement is also true cf. [Ja], but we do not reprove this here.

5.18 Goldie ranks of primitive ideals.

We may now use corollary 5.17.3 to reformulate theorem 5.16 as follows:

Proposition: $\text{rk } H_w^{(\lambda,\mu)} / \text{rk } H_w^{(0,0)} = a! \, \text{rk } H_w^{(0,0)} \, P_{w^{-1}}(\mu+\rho)$

for all $\lambda,\mu \in \Omega$, $w \in W$.

The significance of this alternative formulation is that $P_{w^{-1}}$ depends only on the left cell of w^{-1} (by cor. 5.12.1), so only on the primitive ideal $J := I(w^{-1}\lambda)$, which is the right annihilator of $H_w^{(\lambda,\mu)}$. Since the same is true for
$$a = a(w) = a(w^{-1}) = \dim X - \tfrac{1}{2} \, d \, (U(\mathfrak{g})/J),$$
let us point out the

Corollary 1: *The ratio of left Goldie ranks of two simple Harish–Chandra bimodules as in the proposition is a function of the right annihilators.*

Let us mention that this ratio is also the ratio of the corresponding left characteristic classes (by corollary 5.11), i.e.
$$P(H_w^{(\lambda,\mu)})/ P(H_w^{(0,0)}) = a! \, \text{rk } H_w^{(0,0)} \, P_{w^{-1}}(\mu+\rho). \qquad (1)$$

Theorem (Joseph [J3]): *For all $w \in W$ there is a positive rational constant c_w such that*
$$\text{rk } U(\mathfrak{g})/I(w.\lambda) = c_w \, \tilde{p}_w(\lambda) \, ,$$
for all $\lambda \in \Omega$.

Proof: We take $v \in W$ in the left cell of w as in 5.12, remarks 1, 2, so $H_v^{(\lambda,\lambda)}$ is isomorphic to the socle of $U(\mathfrak{g})/I(w.\lambda)$, and $v^2 = 1$. The right annihilators of these modules are $I(v^{-1}.\lambda) = I(w.\lambda)$ (by 5.12(2)), so

$$P_w = P_{v^{-1}} = P_v \tag{2}$$

by corollary 5.12.1 and $v^{-1} = v$. Hence 5.12(5) gives us

$$\text{rk } U(\mathfrak{g})/I(w.\lambda)/\text{rk } U(\mathfrak{g})/I(w.o) = \text{rk } H_v^{(\lambda,\lambda)} / \text{rk } H_v^{(0,0)} =$$

$$a! \ \text{rk } H_w^{(0,0)} \ P_{v^{-1}} \ (\lambda+\rho) = a! \ \text{rk } H_w^{(0,0)} \ P_w(\lambda+\rho), \tag{3}$$

using $a(v) = a(w) = a$, and the above proposition. Now the theorem follows by using again cor. 5.17.3. $\hspace{3cm}$ Q.e.d.

Corollaries: $\hspace{2cm} P(U(\mathfrak{g})/I(w.\lambda)) = c_w \tilde{p}_w(\lambda) \ P_w, \tag{4}$

$$\mathit{Ch} \ (U(\mathfrak{g})/I(w.\lambda)) = c_w' \ \tilde{p}_w(\lambda) \ [G \times^B V(L(w^{-1}.0))] \tag{5}$$

$$= c_w'' \ [G \times^B V (L(w^{-1}.\lambda))] \tag{6}$$

with positive rational factors c_w, c_w', c_w'' independent of λ.

BIBLIOGRAPHY

[AL]	Alvis, D., Lusztig, G.:	On Springer's correspondence for simple groups of type $E_n (n = 6,7,8)$, Math. Proc. Cambridge Philos. Soc. $\underline{92}$ (1982), 65–78.
[A1]	Atiyah, M.F.:	K–theory; Benjamin, New York–Amsterdam 1967.
[A2]	Atiyah, M.F.:	Characters and cohomology of finite groups; Publ. Math. IHES $\underline{9}$ (1961), 23–64.
[A3]	Atiyah, M.F.:	Power operations in K–theory; Quarterly J. Math. (Oxford) $\underline{17}$ (1966), 165–193.
[AH]	Atiyah, M.F.– Hirzebruch, F.:	Vector bundles and homogeneous space; Diff. Geometry. Proc. of Symp. Pure Math. AMS $\underline{3}$ (1961), 7–38.
[AM]	Atiyah, M.F.– MacDonald, I.G.:	Introduction to commutative algebra; Addison–Wesley (1969).
[AS]	Atiyah, M.F.– Segal, G.B.:	Equivariant K–theory and completion; J. Differential Geometry $\underline{3}$ (1969) 1–18.
[BV1]	Barbasch, D.– Vogan, D.:	Primitive ideals and orbital integrals in complex classical groups; Math. Ann. $\underline{259}$, 153–199 (1982).
[BV2]	Barbasch, D.–	Primitive ideals and orbital integrals in complex exceptional groups; J. Algebra $\underline{80}$ (1983), 350–382.
[BV3]	Barbasch, D. Vogan, D.:	Unipotent representations of complex semisimple groups; Annals of Math. $\underline{121}$ (1985), 41–110.
[BFM]	Baum, P.– Fulton, W.– MacPherson, R.:	Riemann–Roch and topological K–theory for singular varieties; Acta Math. $\underline{143}$ (1979), 155–192.
[BeBe]	Beilinson, A. Bernstein, J.:	Localisation de g–modules; C.R.Acad. Sci. Paris $\underline{292}$, 15–18 (1981).
[BBD]	Beilinson, A.A.– Bernstein, J.– Deligne, P.:	Faisceaux Pervers; Soc. Math. de France, Astérisque 100 (1983).
[BeGe]	Bernstein, J. – Gelfand, S.I.:	Tensor products of finite and infinite dimensional representations of semisimple Lie algebras; Compositio Math. $\underline{41}$, 245–285 (1980).

[SGA6] Berthelot, P.– Théorie des intersections et théorème de
Grothendieck, A. Riemann–Roch; Sém. Geom. Alg. du Bois
Illusie, L.: Marie 1966/67, Lecture Notes in Math. 225,
Springer (1971).

[BS] Beynon, W.M., Green functions of finite Chevalley groups of
Spaltenstein, N.: type E_n (n=6,7,8). J. Algebra 88 (1984),
584–614.

[Bo] Borel, A.: Topics in the homology theory of fibre
bundles; Lecture Notes in Mathematics 36,
Springer (1967).

[Bo2] Borel, A.: Seminar on transformation groups; Annals of
Math. Studies, 46, Princeton (1960).

[BS] Borel, A.– Le Théorème de Riemann–Roch; Bull. Soc.
Serre,J.–P.: Math. France 86 (1958), 97–136.

[B1] Borho, W.: A survey on enveloping algebras of semisimple
Lie algebras; Canad. Math. Soc. Conf. Proc. 5
(1986), 19–50.

[B2] Borho, W.: Recent advances in enveloping algebras of
semisimple Lie Algebras; Sém. Bourbaki 1976,
Springer Lect. Notes in Math. 677, 1–19
(1978)

[B3] Borho, W.: On the Joseph–Small additivity principle for
Goldie–rank, Compositio Math. 47 (1982),
3–29.

[B4] Borho, W.: Definition einer Dixmier–Abbildung für sl
(n,\mathbb{C}); Invent. Math. 40 (1977), 143–169.

[BB1] Borho, W.– Differential operators on homogeneous spaces
Brylinski, J.–L.: I; Invent. Math. 69 (1982), 437–476.

[BB3] Borho, W.– Differential operators on homogeneous spaces
Brylinski, J.–L.: III; Invent. Math. 80 (1985) 1–68.

[BBM1] Borho, W.– Open problems in algebraic groups. (R.
Brylinski, J.–L.– Hotta, N. Kawanaka, eds.) The Taniguchi
MacPherson, R.: Foundation, Katata (Japan) 1983.

[BBM2] Borho, W.– A note on primitive ideals and characteristic
Brylinski, J.–L.– classes; in: Geometry Today (E. Arbarello,
MacPherson, R.: C. Procesi, E. Strickland, eds.), pp. 11–20,
Birkhäuser: Boston – Basel – Stuttgart
(1985).

126

[BBM3]	Borho, W.– Brylinski, J.–L.– MacPherson, R.:	Springer's Weyl Group Representations Through Characteristic Classes of Cone Bundles; Math. Ann. 278, 273–289 (1987).
[BGR]	Borho, W.– Gabriel, R.– Rentschler, R.:	Primideale in Einhüllenden Auflösbarer Lie–Algebren; Lecture Notes in Math. 357, Springer (1973).
[BJ]	Borho, W. Jantzen, J.C.:	Über primitive Ideale in der Einhüllenden einer halbeinfachen Lie–Algebra; Inv. Math. 39 (1977), 1–53.
[BK]	Borho, W.– Kraft, H.:	Uber Bahnen und deren Deformationen bei linearen Aktionen reduktiver Gruppen; Comment. Math. Helvetici 54 (1979), 61–104.
[BK2]	Borho, W. – Kraft, H.:	Uber die Gelfand–Kirillov–Dimension; Math. Annalen 220, 1–24 (1976).
[BM1]	Borho, W.– MacPherson, R.:	Représentations des groupes de Weyl et homologie d'intersection pour les variétés nilpotentes; C.R.Acad. Sci. Paris (A) 292 (1981), 707–710.
[BM2]	Borho, W.– MacPherson, R.:	Partial resolutions of nilpotent varieties; in: Analyse et Topologie sur les Espaces Singuliers, Soc. Math. de France, Astérisque 101 (1983), 23–74.
[AC]	Bourbaki, N.:	Algèbre commutative: Paris – New York – Masson, 1983.
[BKa]	Brylinski, J.–L. Kashiwara, M.:	Kazhdan–Lusztig conjecture and holonomic systems; Invent. Math. 64, 387–410 (1981).
[D1]	Deligne, P.:	Théorie de Hodge III, Publ. Math. IHES 4 (1974), 6–77.
[Di]	Dixmier, J.:	Algèbres enveloppantes; Paris: Gauthier Villars, 1974.
[D]	Duflo, M.:	Sur la classification des idéaux primitifs dans l'algebre enveloppante d'une algèbre de Lie semisimple; Ann. Math. 105, 107–120 (1977).
[D2]	Duflo, M.:	Théorie de Mackey pour les groupes de Lie algébrique; Acta Math. 149 (1982), 153–213.
[E]	Enright, T.J.:	Lectures on representations of complex semisimple Lie groups; Tata Institute Lectures on Math., Berlin et al.: Springer, 1981.

[Fr]	Friedlander:	Etale homotopy of simplicial schemes; Annals of Math. Studies 104, Princeton, 1982.
[Fu]	Fulton, W.:	Intersection Theory; Springer: Berlin – Heidelberg – New York – Tokyo, 1984.
[FL]	Fulton, W.– Lang, S.:	Riemann–Roch algebra; Springer: Berlin – Heidelberg – New York – Tokyo, 1984.
[FM]	Fulton, W.– MacPherson, R.:	Intersecting cycles on an algebraic variety; in: Real and complex singularities, Oslo 1976 (P. Holm ed.), Sijthoff & Noordhoff: Alphen a.d. Rijn, 1977.
[Gi]	Ginsburg, V.:	G–modules, Springer's representations and bivariant Chern classes; Adv. in Math 61 (1986), 1–48.
[Gi2]	Ginsburg, V.:	Deligne–Langlands conjecture and representations of affine Hecke algebras, preprint, Moscow, 1985
[Gi3]	Ginsburg, V.:	Fourier–Langlands transformations on reductive groups, Funct. Anal. and its Appl. 22 (1988) 143–44.
[Gd]	Godement, R.:	Topologie Algébrique et Théorie des Faisceaux; Paris: Hermann, 1958.
[Go]	Goldie, A.W.:	The structure of noetherian rings, in: Lectures on rings and modules, Springer Lecture Notes in Math. 246, 213–321 (1972).
[GM1]	Goresky, M.– MacPherson, R.:	Intersection Homology Theory; Topology 19 (1980), 135–162.
[GM2]	Goresky, M.– MacPherson, R.:	Intersection Homology II; Invent. Math. 72 (1983), 77–130.
[Ha]	Hadziev, D.:	Some questions in the theory of vector invariants; Math. USSR – Sb. 1 (1967), 383–396.
[H1]	Hiller, H.:	Geometry of Coxeter groups; Res. Notes in Math. 54, Pitman: Boston – London – Melbourne, 1982.
[Hi]	Hirzebruch, F.:	Topological methods in algebraic geometry; Springer: Berlin – Heidelberg, 1956.
[Ho]	Hotta, R.:	On Joseph's construction of Weyl group representations; Tohoku Math. J. 36 (1984), 49–74.

[H2] Hotta, R.: On Springer's representations, J. Fac. Sci.,
 Univ. of Tokyo, IA 28 (1982), 863–876.

[H3] Hotta, R.: Character \mathcal{D}–modules on a reductive group;
 preprint. Sendai, 1987.

[HK] Hotta, R. – The invariant holonomic system on a
 Kashiwara, M.: semisimple Lie algebra; Invent. Math. 75,
 327–358 (1984).

[Hu] Humphreys, J.E.: Introduction to Lie Algebras and
 Representation Theory; New York –
 Heidelberg – Berlin; Springer 1972.

[Ja] Jantzen, J.C.: Einhüllende Algebren halbeinfacher
 Lie–Algebren; Springer: Berlin–Heidelberg–
 New York – Tokyo, 1983.

[J1] Joseph, A.: On the variety of a highest weight module; J.
 of Algebra 88 (1984), 238–278.

[J2] Joseph, A.: On the associated variety of a primitive ideal;
 J. of Algebra 93 (1985), 509–523.

[J3] Joseph, A.: Goldie rank in the enveloping algebra of a
 semisimple Lie algebra I, II; J. of Algebra 65
 (1980), 269–306.

[J4] Joseph, A.: Dixmier's problem for Verma– and principal
 series modules; J. London Math. Soc. (2) 20,
 193–204 (1979).

[J5] Joseph, A.: On the annihilators of the simple subquotients
 of the principal series; Ann. Sci. E.N.S. (4) 10,
 419–440 (1977).

[J6] Joseph, A.: A surjectivity theorem for rigid highest
 weight modules; preprint Rehovot and Paris,
 1987.

[KT] Kashiwara, M. – The characteristic cycles of holonomic systems
 Tanisaki, T.: on a flag manifold; Invent. Math. 77, 185–198
 (1984).

[KL] Kazhdan, D.A. – Representations of Coxeter groups and Hecke
 Lusztig, G.: algebras; Invent. Math. 53, 165–184 (1979).

[KL2] Kazhdan, D., A topological approach to Springer's
 Lusztig, G.: representations; Advances in Math. 38 (1980),
 222–228.

[KL3] Kazhdan, D., Proof of the Deligne–Langlands conjecture for
 Lusztig, G.: Hecke algebras, Inventiones 87 (1987),
 153–215

[Ki] King, D.R.: The character polynomial of the annihilator of an irreducible Harish–Chandra module; Amer. J. Math. <u>103</u> (1981), 1195–1240.

[Kn] Knutson, P.: Lambda–rings and the representation ring of the symmetric group; Lecture Notes in Math. <u>303</u>, Springer, 1973.

[Kr] Kraft, H.: Geometrische Methoden in der Invariantentheorie; Braunschweig-Wiesbaden: Vieweg, 1984.

[Le] Levasseur, T.: Equidimensionalité de la variété caractérisque; (after O. Gabber) manuscript, Paris, 1982.

[LSS] Levasseur, T. Smith, S.P. Stafford, J.T.: The minimal nilpotent orbit, the Joseph ideal and differential operators; preprint, 1987.

[L] Lusztig, G.: A class of irreducible representations of a Weyl group; Proc. Kon. Nederl. Akad. Wetensch. <u>82</u> (1979), 323–335.

[Lu] Lusztig, G.: Green polynomials and singularities of unipotent classes; Adv. in Math. <u>42</u> (1981), 169–178.

[Ma] Manin, Y.: Lectures on the K–functor in algebraic geometry; Russ. Math. Survey <u>24</u> (1969), 1–89.

[Mr] Marlin, R.: Thése; Nice, 1979.

[MG] McGovern, W.M.: Unipotent representations and Dixmier algebras; preprint, Berkeley, 1987.

[Mi] Miličić, D.: Localization and representation theory of reductive Lie groups; preprint, Salt Lake City, 1987.

[MR] Moeglin, C.: Sur la classification des idéaux primitifs des algèbres enveloppantes; Bull. Soc. Math. France <u>112</u> (1984), 3–40.

[Me] Melnikov, A. Thesis, Weizman Institute (Rehovost).

[M1] Moeglin, C.: Idéaux complètement premiers de l'algèbre enveloppanté de $gl_n(\mathbb{C})$; preprint, Paris, 1985.

[M2] Moeglin, C.: Modèles de Whittaker et idéaux primitifs complètement premiers dans les algèbres enveloppantes, I resp. II; preprints, Paris, 1986.

[M3] Moeglin Sur la classification des idéaux primitifs des algébres enveloppantes; Bull. Soc. Math. France 112 (1984), 3–40.

[Mu] Mumford, D.: Geometric Invariant Theory; Heidelberg: Springer, 1965.

[Ro] Rosenlicht, M.: Some basic theorems on algebraic groups; Amer. J. Math. $\underline{78}$ (1956), 401–443.

[Se1] Segal, G.: Equivariant K–theory; Publ. Math. IHES $\underline{34}$ (1968), 129–151.

[Se2] Segal, G.: The representation ring of a compact Lie group; Publ. Math. IHES $\underline{34}$ (1968), 113–128.

[Sr] Serre, J.–P.: Algèbre locale–Multiplicités; Lecture Notes in Math. $\underline{11}$, Springer, 1965.

[Sr2] Serre, J.–P. Espaces fibres algebriques; Seminaire C. Chevalley, E.N.S., avril 1958 (pp. 1–01 to 1–37).

[Sh] Seshadri, C.S.: Geometric reductivity over arbitrary base; Advances in Math. $\underline{26}$ (1977), 225–274.

[Sh1] Shoji, T.: On the Springer representations of the Weyl groups of classical algebraic groups; Comm. Algebra 7 (1979), 1713–1745; correction in Comm. Algebra $\underline{7}$ (1979), 2027–2033.

[Sh2] Shoji, T.: On the Springer representations of Chevalley groups of type F_4; Comm. Algebra $\underline{8}$ (1980), 409–440.

[Sl] Slodowy, P.: Four lectures on simple groups and singularities; Comm. Math. Inst. Utrecht, 1980.

[Sm] Smoke, W.: Dimension and multiplicity for graded algebras; J. Algebra $\underline{210}$ (1972), 149–173.

[Sp] Spaltenstein, N.: On the generalized Springer correspondence for exceptional groups; Advanced Studies in Pure Math. $\underline{6}$, Algebraic groups and related topics (R. Hotta, ed.); North–Holland: Amsterdam – New York – Oxford, 1985.

[Sp2] Spaltenstein, N.: On the fixed point set of a unipotent element on the variety of Borel subgroups; Topology 16 (1977), 203–204.

[S1] Springer, T.A.: Trigonometric sums, Green functions of finite groups and representations of Weyl groups; Invent. Math. 36 (1976), 173–207.

[S2] Springer, T.A.: A construction of representations of Weyl groups, Invent. Math. 44 (1978), 279–293.

[S2] Springer, T.A.: Linear algebraic groups; Boston – Basel – Stuttgart: Birkhäuser, 1981.

[St] Steinberg, R.: On the desingularization of the unipotent variety; Invent. Math. 36 (1976), 209–224.

[St2] Steinberg, R.: Conjugacy classes in algebraic groups; Lecture Notes in Math. 366, Berlin – Heidelberg – New York: Springer, 1974.

[Ta] Tanisaki, T.: Associated varieties of highest weight modules; manuscript, Tohoku Univ., Sendai, 1985.

[T1] Thomason, R.W.: Algebraic K–theory of group scheme actions; preprint, Baltimore, 1985 (to appear in Annals of Math. Studies).

[T2] Thomason, R. W.: Comparison of equivariant algebraic and topological K–theory; preprint, Johns Hopkins Univ., Baltimore, 1985.

[V] Vogan, D.: Ordering of the primitive spectrum of a semisimple Lie algebra; Math. Ann. 248, 195–203 (1980).

[V2] Vogan, D.: The orbit method and primitive ideals for semisimple Lie algebras; Canadian MS Conference proceedings 5 (1986), AMS, Providence 1986.

Progress in Mathematics

Edited by:

J. Oesterlé
Departement des Mathematiques
Université de Paris VI
4, Place Jussieu
75230 Paris Cedex 05
France

A. Weinstein
Department of Mathematics
University of California
Berkeley, CA 94720
U.S.A.

Progress in Mathematics is a series of books intended for professional mathematicians and scientists, encompassing all areas of pure mathematics. This distinguished series, which began in 1979, includes authored monographs and edited collections of papers on important research developments as well as expositions of particular subject areas.

All books in the series are "camera-ready", that is they are photographically reproduced and printed directly from a final-edited manuscript that has been prepared by the author. Manuscripts should be no less than 100 and preferably no more than 500 pages.

Proposals should be sent directly to the editors or to: Birkhäuser Boston, 675 Massachusetts Avenue, Suite 601, Cambridge, MA 02139, U.S.A.

A complete list of titles in this series is available from the publisher.